national STATISTICS

Olga Evans
Nicola Singleton
Howard Meltzer
Robert Stewart
Martin Prince

AF838405
UNIVERSITY

Morbidity among Adults in Great Britain carried out in 2000 for the Department of Health, the Scottish Executive Health Department and the Welsh Assembly Government

The Mental Health of Older People

London: TSO

© Crown copyright 2003
Published with the permission of the Controller of Her Majesty's Stationery Office (HMSO).

ISBN 0 11 621660 3

Applications for reproduction should be submitted to HMSO under HMSO's Class Licence:
www.clickanduse.hmso.gov.uk

Alternatively applications can be made in writing to:

HMSO
Licensing Division
St. Clement's House
2–16 Colegate
Norwich
NR3 1BQ

Contact points

For enquiries about this publication, contact
Nicola Singleton
Tel: **020 7533 5305**
E-mail: **nicola.singleton.@ons.gov.uk**

To order this publication, call TSO
on **0870 600 5522.** See also back cover.

For general enquiries, contact the National Statistics
Customer Enquiry Centre on **0845 601 3034**
(minicom: 01633 812399)
E-mail: **info@statistics.gov.uk**
Fax: 01633 652747
Letters: Room D115, Government Buildings,
Cardiff Road, Newport NP10 8XG

You can also find National Statistics on the Internet
at **www.statistics.gov.uk**

About the Office for National Statistics

The Office for National Statistics (ONS) is the government agency responsible for compiling, analysing and disseminating many of the United Kingdom's economic, social and demographic statistics, including the retail prices index, trade figures and labour market data, as well as the periodic census of the population and health statistics. The Director of ONS is also the National Statistician and the Registrar General for England and Wales, and the agency that administers the registration of births, marriages and deaths there.

This report has been produced by the Social Survey Division of the Office for National Statistics in accordance with the National Statistics Code of Practice.

MANCHESTER
METROPOLITAN
-4 JUN 2004
UNIVERSITY LIBRARY

Contents

List of tables

List of figures

Authors' acknowledgements

We would like to thank everybody who contributed to the survey and the production of this report. We were supported by our specialist colleagues in ONS who carried out the sampling, field work and computing elements for the survey.

Great thanks are also due to all the ONS interviewers who worked on the survey.

We were assisted at all stages of the survey by a group of expert advisors who we would like to thank for the valuable specialist advice they provided. The group comprised:

Professor P Bebbington, University College, London
Professor T Brugha, University of Leicester
Dr D Bughra, Institute of Psychiatry, London
Professor J Coid, Forensic Psychiatry Research Unit, St Bartholemew's Hospital
Dr M Farrell, Institute of Psychiatry, London
Professor G Lewis, University of Wales, Cardiff
Dr M Prince, Institute of Psychiatry, London

The project was steered by a group comprising the following, to whom thanks are also due for assistance and advice given at various stages of the survey.

Ms J Davies (chair), Department of Health
Mr J O'Shea (secretariat) , Department of Health
Mr R Bond, Department of Health
Mr A Boucher, Department of Health
Ms S Carey, Office for National Statistics
Mr D Daniel, Department of Health
Dr T Fryers, University of Leicester
Dr S Gupta, Department of Health
Dr A Higgitt, Department of Health
Professor R Jenkins, Institute of Psychiatry, London
Dr D Jones, Department of Health
Ms T Jones, National Assembly for Wales
Dr J Loudon, Scottish Executive
Mr G Russell, Scottish Executive
Mr J Sweeney, National Assembly for Wales

Most importantly, we would like to thank all the participants in the survey for their time and co-operation.

Notes to tables

1 Tables showing percentages

The row or column percentages may add to 99% or 101% because of rounding.

The varying positions of the percentage signs and bases in the tables denote the presentation of different types of information. Where there is a percentage sign at the head of a column and the base at the foot, the whole distribution is presented and the individual percentages add to between 99% and 101%. Where there is no percentage sign in the table and a note above the figures, the figures refer to the proportion of people who had the attribute being discussed, and the complementary proportion, to add to 100%, is not shown in the table. If the bases are shown at the bottom of the table the figures shown are column percentages, i.e. they reflect the percentage of people within the category described at the head of the column who had the attribute under consideration. In other tables the bases are shown on the right hand side and the note above the figures describes the attribute under consideration and in this case the figures shown are the percentage of people with that attribute in the group described by combining the column and row headings. For example in Table 3.2 the percentage of respondents who had a CIS-R score of 12 or above among men who had experienced none or one stressful life event was 5%, while for men who had experienced two such events it was 2%.

The following conventions have been used within tables:

-	no cases
0	values less than 0.5%
..	data not available

2 Statistical significance

Unless otherwise stated, differences mentioned in the text have been found to be statistically significant at the 95% confidence level. Standard errors that reflect the complex sampling design and weighting procedures used in the survey have been calculated and used in tests of statistical significance.

3 Small bases

Very small bases have been avoided wherever possible because of the relatively high sampling errors that attach to small numbers. In general, percentage distributions are shown if the base is 30 or more. Where the base is lower, actual numbers are shown in square brackets.

Summary of key findings

1 Background and methods used

- This report is one of several topic reports looking at data from the survey carried out by ONS on psychiatric morbidity of adults aged 16–74 living in private households in Great Britain. This report focuses on the prevalence of neurotic symptoms and disorders, and the distribution of cognitive function, in adults aged 60–74 years. It also examines variations in prevalence associated with a wide range of social and economic factors, and the interaction between neurotic symptoms, cognitive functioning, and physical illness.

- The survey on which this report is based was commissioned by the Department of Health, the Scottish Executive Health Department and the Welsh Assembly Government.

- The survey was carried out in 2000 and is a repeat of the 1993 survey of adults living in private households. However, in the 2000 survey there was a slight increase in the age range, so that it covered people aged 16 up to 74 years, compared with 16 to 64 in the 1993 survey. The topics covered were also expanded. Most pertinent to this report was the addition in 2000 of the assessment of cognitive function in adults aged 60 years and over, and an expansion of the indices of economic and social functioning covered.

- A chapter of the report is dedicated to each of the following areas:
  - the prevalence of common mental disorder and variations in prevalence with socio-economic characteristics;
  - variations in prevalence of common mental disorder with experience of stressful life events, social support, and physical health problems;
  - the impact of common mental disorder on self-reported mental and physical well-being, activities of daily living and service use;
  - the distribution of cognitive function and variations in distribution with socio-economic characteristics;
  - the impact of cognitive function on self-reported mental and physical well-being, activities of daily living and service use; and
  - the association between cognitive impairment and common mental disorder and physical health problems.

2 Prevalence of common mental disorder

- One in ten respondents (10%) had significant levels of neurotic symptoms (CIS-R scores of 12 or more). Half of these (5% of the sample as a whole) reported levels of symptoms of a severity equivalent to that found among people being treated in secondary care (CIS-R scores of 18 or more). About one in eight (12%) respondents had at least one neurotic disorder.

- Women were more likely than men to have significant levels of neurotic symptoms (12% of women compared with 8% of men), and to have any neurotic disorder (14% compared with 9%).

- The prevalence of common mental disorder decreased with age, but this was much more pronounced in men than in women, and statistically significant only in men. Among men 13% of those aged 60–64 years, compared with just 5% of those aged 65 to 69 years, and 6% of those aged 70–74 years had CIS-R scores of 12 or more. Among the women the corresponding figures for each age group were 13%, 12%, and 11% respectively.

- People who were divorced or separated had the highest prevalence of common mental disorder (19% of women and 17% of men) and being single was associated with a low prevalence in both sexes (4% of women and 8% of men). However, while being married was associated with a low prevalence of mental disorder in men (7%) married women had a higher prevalence (12%).

- Common mental disorder was more prevalent in the manual social classes (III manual, IV, and V) than in the non-manual social classes (I or II combined, III non-manual). The prevalence was highest in Social Class V (18% with CIS-R scores of 12 or more) and lowest in Social Classes I or II combined (6%).

- The prevalence of common mental disorder increased steadily with decreasing household income. While just one in fifty (2%) of those with an income of £500 or more had a CIS-R score of 12 or more the prevalence increased to about one in seven respondents (15%) with a weekly household income below £200.

- Being in receipt of the types of state benefit that are either means tested, or contingent upon disability, was associated with a greatly increased likelihood of having a mental health problem. At least one out of four respondents in receipt of any disability related benefit (27%), any income related benefit (29%), or housing benefit (24%) had CIS-R scores of 12 or more, making them at least twice as likely as respondents in general (10%) to have a high CIS-R score.

- Of the socio-economic variables discussed above, six were found to be independently associated with having clinically significant levels of neurotic symptoms. These were: sex, social class, household income, receipt of a disability related benefit, receipt of an income related benefit, and receipt of any other state benefit (excluding housing benefit).

3 Other factors associated with common mental disorder

- The number of stressful life events ever experienced was positively associated with prevalence of mental health problems. Only around one in twenty respondents reporting one (5%) or two (6%) life events had a CIS-R score of 12 or more, but this rose to one in eight (13%) of those reporting five life events, and to one in five (20%) of those reporting six or more life events.

- People were about three times more likely to have mental health problems if they had ever experienced a financial crisis than if they had not (29% compared with 9%), and about twice as likely to have mental health problems if they had ever experienced serious illness, injury or assault to themselves (14% compared with 7%), or if they had ever experienced separation or divorce (18% compared with 9%), than if they had not.

- Fewer than one in ten respondents (9%) who had not experienced a single stressful life event in the six months prior to interview, compared with one in seven respondents (14%) who had experienced one event, and one in four respondents (25%) who had experienced two or more events in the preceding six months had a CIS-R score of 12 or more.

- People with a longstanding physical health problem were more likely than those without to have a mental health problem and the likelihood increased with increasing extent of physical health problems. Just 3% of respondents with no physical health problems had a CIS-R score of 12 or more, compared with 11% of those with one organ system affected, 14% of those with two and 23% of respondents with three or more organ systems affected.

4 Impact of common mental disorder

- The SF-12 is a self-report measure of health status that is scored to generate two summary scores – one for mental health, (the MCS-12), and one for physical health, (the PCS-12). Potential scores on the MCS-12 and PCS-12 range from 0–100, with higher scores indicating better health.

- Women had slightly lower MCS-12 scores than men but there was no statistically significant difference in PCS-12 scores. The mean MCS-12 score for women was 51.2, and for men was 53.1, while the mean physical health summary score was 44.8 for women and 45.3 for men.

- Mean MCS-12 and PCS-12 scores were lower among people with mental health problems as indicated by a CIS-R score of 12 or more than among those without mental health problems. The mean MCS-12 score was 38.7 for respondents with mental health problems, and 53.6 for those without mental health problems. The corresponding values of the PCS-12 score were 33.4 and 46.4.

- Mean SF-12 mental and physical health component scores decreased significantly with increasing levels of physical health problems. However, at every level of physical ill-health (number of organ systems affected), mean MCS-12 scores were lower among respondents with mental health problems as shown by a CIS-R score of 12 or above than among those without. For example, among respondents with one organ system affected by long-term physical problems, mean MCS-12 scores were 38.6 for those with a CIS-R score of 12 or more compared with 53.6 for those with lower CIS-R scores.

- Consistent with the information on variation in distribution of neurotic symptoms and neurotic disorders with age, people aged 60–64 had lower mean MCS-12 scores than older respondents. Given the association between age and decline in physical health, it was not surprising that, in contrast, mean PCS-12 scores were highest in the 60–64 age group and lowest in the 70–74 year age group. However, both mean MCS-12 and PCS-12 scores were lower among respondents with mental health problems at every age group. For example, in the 60–64 year age group respondents with a CIS-R score below 12 had a mean MCS-12 score of 53.1, while those with a CIS-R score of 12 or more had a mean score of 38.9. The corresponding mean PCS-12 scores were 48.1 compared with 34.0.

- Over a third of all respondents (37%) reported having difficulty with one or more of the seven activities of daily living (ADLs) covered in this survey. Almost half of these respondents (18% overall) reported difficulties with just one ADL.

- The likelihood of reporting difficulties with at least one ADL rose steadily with increasing CIS-R score. Respondents with few, if any, neurotic symptoms (CIS-R scores 0 to 5) were very much less likely than any other respondents to have difficulty with any ADL: 28% compared with 60% of respondents with CIS-R scores between 6 and 11, 70% of respondents with scores between 12 and 17, and 78% of respondents with scores of 18 and over.

- As expected, likelihood of reporting difficulty with any ADL increased with increasing extent of long-term physical health problems. However, at every level of physical ill-health respondents with mental disorder were more likely than those without to have difficulty with at least one ADL. For example, among respondents with one organ system affected by physical illness, 35% of those with a CIS-R score below 12 compared with 66% of those with a CIS-R score of 12 or more had difficulties with at least one ADL. Among respondents with three or more organ systems affected by physical illness the corresponding figures were 65% compared with 94%.

- At each age group, respondents with mental disorder were more likely than those without to have difficulty with at least one ADL. For example, among respondents aged 60–64, 27% of those with a CIS-R score below 12 compared with 64% of those with a CIS-R score of 12 or more had difficulties with at least one ADL. Among respondents aged 65– 69 the corresponding figures were 34% compared with 83%, and among those aged 70– 74, 41% compared with 81%.

- Across all respondents, 'Practical activities' was the ADL most frequently associated with difficulty (29% of respondents had difficulty with this). The next most common sources of difficulties were 'dealing with paperwork' (14%), 'household activities' (13%), and 'using transport' (13%). Almost one in ten respondents (9%) had difficulties with 'personal care', but just 3% had difficulties 'managing money' and just 2% had difficulties with 'medical care'.

- Respondents with mental health problems were more likely than respondents without to have difficulty with every individual ADL. For example, those with CIS-R scores of 12 or above were around five times as likely as those with lower CIS-R scores to report difficulties with: household activities (44% compared with 9%); managing money (10% compared with 2%); personal care (29% compared with 6%); medical care (6% compared with 1%).

- People were asked about their use of a range of services for both physical and mental health problems. The General Practitioner service was the most commonly used community based service. Almost three-quarters of all respondents (72%) had visited their GP with a physical complaint in the year before interview, and about one in twelve (8%) had visited with a mental complaint. Only 6% of all respondents had used any other community care in the year prior to interview.

- Respondents with mental health problems were more likely than those without to use most services, whether the consultations were ostensibly for physical or mental health problems. People with CIS-R scores of 12 or more were more likely than those with lower CIS-R scores to have consulted a GP about a physical complaint in the year prior to interview, (90% compared with 70%), and more likely to have consulted a GP about a mental complaint, (30% compared with 5%). They were also three times as likely as people with lower CIS-R scores to have received some other form of community care, (15% compared with 5%).

- A quarter of all respondents (25%) had had an outpatient or day-patient hospital visit in the quarter preceding the interview, but this had been for a mental complaint in fewer than one percent of cases. Only one in twenty-five of all respondents (4%) had had an inpatient hospital stay in the quarter preceding the interview, and again this had been for a mental complaint in fewer than one percent of cases.

- People with mental health problems were more likely than those without such problems to have had hospital outpatient (44% compared with 23%) or inpatient treatment (9% compared with 4%) for physical health problems in the quarter prior to interview.

- At every level of physical health problems, people assessed as having a mental health problem at the time of interview were more likely than those without such problems to have consulted their GP about a mental or emotional complaint and physical health problems. They were also more likely to have used other services.

- At every age group, a higher proportion of people with mental health problems than those without had consulted a GP for mental and for physical complaints, had out-patient or in-patient treatment and to have used any other community care services although in the higher age groups the differences did not all reach statistical significance.

5 Distribution of cognitive function

- Cognitive function was assessed using the following three measures:
  - The TICS-m: a short measure to screen for dementia.
  - The Animal Naming Test: a test of verbal fluency (VF).
  - The National Adult Reading Test (NART): a measure of crystallised intelligence.

- About one out of five (19%) of all respondents had a TICS-m score below 21 (indicative of cognitive impairment), and there was no difference between men and women in likelihood of having a low TICS-m score. Across all respondents, mean TICS-m total score was 23.7, with no sex difference in scores.

- Among both men and women, the likelihood of having a TICS-m score below 21 increased with age. Among men, the likelihood of having a low TICS-m score did not differ between the 60 to 64 years and 65 to 69 years age groups (16% in each age group had low TICS-m scores), but it was significantly higher among those aged 70 to 74 years (25% had low TICS-m scores) than in the younger age groups. Among women, the increase in the proportion with cognitive impairment occurred at an earlier age. About one out of seven women (14%) aged 60 to 64 years had a total TICS-m score below 21 but this increased to over one out of five women (22%) aged 65–69 years, and one in four women (26%) aged 70–74 years.

- TICS-m scores were strongly associated with the level of educational qualifications attained in both sexes. The likelihood of having a score below 21 decreased with increasing level of education.

- The proportion of people who had a TICS-m score below 21 tended to increase with decreasing social class, with the difference between non-manual and manual social classes being the most marked. While 10% of people in Social Classes I and II and 14% of those in class III non-manual showed signs of cognitive impairment this rose to 23% of those in class III manual, 26% of those in IV and 36% of those in Social Class V.

- The likelihood of having TICS-m scores indicative of cognitive impairment decreased with increasing household income in both men and women. More than 1 in 4 people with a gross weekly household income of less than £200 had a TICS-m score below 21, and this decreased to less than 1 in 14 of those with a weekly income £400 or more.

- Both men and women whose only source of income was state benefits were more likely than other people to have a TICS-m score below 21, (30% compared with 17% of respondents with income from benefit plus other sources, and 12% of respondents with non-benefit income only).

- Logistic regression analysis indicated that four socio-economic variables were independently associated with having TICS-m scores indicative of cognitive impairment. These were: age, level of highest educational qualifications, social class, and household income.

- The mean NART score for the sample as a whole was 27.8, and did not differ between men and women, nor did it vary with age.

- As might be expected, mean NART scores increased with level of educational qualifications attained. The mean NART score was 22.7 among respondents with no qualifications, this rose to 31.0 among respondents with GCSE level qualifications, and to 37.8 among respondents with qualifications at 'A' level or above.

- People in the non-manual Social Classes (I or II, and III non-manual) had higher mean NART scores than people in the manual social classes (III manual, IV, and V). The mean NART score for Social Classes I or II combined was 35.0 and mean scores declined to 19.2 in Social Class V.

- Mean NART scores also increased with increasing weekly household income. For example, mean NART scores were 36.2 for people with a weekly household income of £500 or more, 30.5 for those with an income between £300 and £399, and 24.9 for those with a weekly household income under £100.

- People in receipt of a disability-related benefit, an income-related benefit, or housing benefit, had significantly lower mean NART scores than respondents in general. Correspondingly, people whose only source of income was state benefits had lower mean NART scores than other people. The mean NART score was 22.2 for those dependent solely on benefits compared with 29.3 among those whose income came from a combination of state benefits and other sources and 29.0 among people whose income came entirely from other sources.

- The mean Verbal Fluency (VF) score (the number of animals named in one minute), across all respondents, was 21.7, with no sex difference in scores. Verbal fluency decreased with increasing age in both men and women. The mean score ranged from 23.0 among people aged 60–64 down to 20.2 for people aged 70–74 years.

- Verbal fluency increased with increasing level of educational qualifications attained. Mean VF scores were 19.9 among people with no educational qualifications and 25.3 among those with qualifications at 'A' level or above.

- Mean VF score tended to increase with social class. People in Social Classes I or II combined had a higher mean VF score than people in Social Classes IV or V, (23.7 compared with 19.9 and 19.6 respectively).

- Verbal fluency was highest among respondents in the two highest weekly household income groups, (between £400 and £499, and £500 and over) and lowest in respondents with a household income between £100 and £199. Respondents with a mean income of £500 or more had a mean VF score of 24.7, and those with a mean income between £100 and £199 had a mean VF score of 19.8.

- People dependent solely on state benefits for their income had a lower mean VF score than did people whose income came wholly or partly from other sources. The mean VF score for those dependent on state benefits was 19.1, this rose to 21.8 among those whose income came from benefits plus other sources, and to 25.3 for those whose income came entirely from other sources.

- The likelihood of having signs of cognitive impairment increased with decreasing NART score. Taking people with all levels of educational qualifications together, those with the lowest NART scores (0 to 19) were about twice as likely as those with a NART score between 20 to 29 to have a TICS-m score below 21, (33% compared with 17%), and more than five times as likely as those with the highest NART scores (40 to 50) to have a low TICS-m score, (33% compared with 6%).

6 Impact of cognitive function

- Although mean MCS-12 and PCS-12 scores tended to be lower in people with evidence of cognitive impairment than those with higher TICS-m scores, at different levels of physical health problems and neurotic symptoms, the differences between people with and without evidence of cognitive impairment were not consistently statistically significant.

- People with evidence of cognitive impairment were more likely than those without to have difficulty with at least one activity of daily living, (52% compared with 34%). This difference was observed at all levels of physical health problems. The effect of cognitive impairment was statistically significant among people with no physical health problems (32% and 12% respectively having difficulty with at least one ADL), and those with just one organ system affected, (53% and 34%, respectively).

- Similarly, the difference in likelihood of having difficulty with at least one ADL between those with and without cognitive impairment was evident among respondents with and without mental health problems, but was only statistically significant among people with CIS-R scores below 12.

- People with signs of cognitive impairment were significantly more likely to have difficulty with each of the seven ADLs asked about, except medical care, than were those with a TICS-m score of 21 or more. For example, people with cognitive impairment were three times as likely as those without to experience difficulty with paperwork, (30% compared with 10%), and four times as likely as those without to report difficulty (8% compared with 2%).

- The likelihood of using health services did not differ significantly between people with and without evidence of cognitive impairment.

7 Comorbidity

- People with TICS-m scores indicative of possible cognitive impairment were more likely than those with higher TICS-m scores to have CIS-R scores indicative of clinically significant levels of neurotic symptoms, although the difference reached statistical significance in men only. About one out of six (16%) of all respondents with a TICS-m score below 21, compared with under one out of ten (9%) of all respondents with a TICS-m score of 21 or more, had a CIS-R score of 12 or over.

- Respondents of both sexes with a TICS-m score below 21 were more likely than those with higher scores to have longstanding physical health problems, but this difference was statistically significant only among women.

1 Background and methods used

1.1 Background

This report is one of several topic reports looking at data from a survey of psychiatric morbidity among adults aged 16–74 living in private households in Great Britain carried out by the Office for National Statistics in 2000. This survey was commissioned by the Department of Health, the Scottish Executive Health Department and the Welsh Assembly Government. This report focuses on the mental health of older people, those aged 60–74 years. Specifically, it examines the distribution of neurotic symptoms such as anxiety and depression, common mental disorders (also known as neurotic disorders), and indices of cognitive functioning among people in that age group.

Other topic reports already available cover:

- suicidal thoughts and behaviours (Meltzer *et al*, 2002a);
- the social and economic circumstances of people with mental disorder (Meltzer *et al*, 2002b);
- people with psychotic disorder (O'Brien *et al*, 2002); and
- tobacco, alcohol and drug use and mental health (Coulthard *et al*, 2002).

This report is based on a survey carried out in 2000 which was a repeat of the 1993 baseline survey of adults living in private households. However, the 2000 survey included a number of developments. There was a slight increase in the age range, so that it covered people aged 16–74 years compared with 16–64 years in the 1993 survey and measures of personality disorder and intellectual functioning were included (Singleton *et al*, 2001). There was also an expansion of the topics covered in relation to economic and social functioning.

Although the main report of the survey (Singleton *et al*, 2001) covered the whole range of psychiatric morbidity across the entire age range (from 16 to 74 years), it did not report on the indices of cognitive functioning that are covered in this report. This report also differs from the main report in its examination of a broader range of social and economic characteristics, and a more restricted range of psychiatric morbidity. In this report only neurotic symptoms and disorders are examined as the low prevalence of psychotic disorder and alcohol and drug dependence among older people, means that the number of survey respondents affected by these conditions would be too small to allow statistical analysis.

1.2 Research context

Common mental disorders

Most previous research in this area has focussed upon depressive disorders in older people ('late-life depression') rather than the somewhat broader category of 'common mental disorders', which would also include the anxiety disorders, panic and phobia, that is neurotic disorders in general. Late-life depression, when defined according to the broad criterion of clinical significance, is a common disorder affecting 10–15% of the over 65 year old population (Beekman *et al*, 1999; Copeland *et al*, 1999). Prevalence rates for 'major depression' (the most severe form of clinical depression) are substantially lower. Longitudinal population-based studies suggest that incidence (new cases) and maintenance rates (persistent cases) are both high, balanced by a high mortality ratio for those affected.

The data on the prevalence of major depression from the United States Epidemiological Catchment Area (ECA) study suggested a lower rate for those over 65 years old (1.0 per cent) than for those aged 45–64 (2.3%) and those aged 18–44 (3.4%) (Weissman *et al*, 1988). This relatively low prevalence rate among the older population was confirmed in a Canadian study using similar methods (Bland *et al*, 1988). These findings have been particularly controversial, as they could be taken to imply that the management of depression may require less resources per capita for the old than for the young (Snowdon, 1990). They conflict with the general impression that the frequency of depressive symptoms and broader depressive syndromes either increases (Tannock and Katona, 1995; Kanowski, 1994; Ernst and Angst, 1995) or remains stable (Henderson *et al*, 1993) with increasing age. The finding of a decreasing prevalence with increasing age had been thought to reflect a bias in the way that the research diagnoses were assigned with respect to older people, in that symptoms that were attributed to bereavement and physical ill health were excluded. Older people report as many lifetime depressive symptoms as younger subjects, but are more likely to attribute them to physical causes, meaning that they are then excluded as a basis for diagnosing depression (Knauper and Wittchen, 1994). The current survey is one of the very few worldwide that has assessed the prevalence of depressive disorders and common mental disorders across the adult life span and into older age, using the same methods to identify mental health problems in different age groups. The bias referred to above should not be a factor, given the methods used in this survey to identify cases of common mental disorder.

Risk factors for late-life depression are in many ways similar to those for younger people (Prince and Beekman, 2001). There is a consistent gender difference, with women affected more than men. Adverse socio-economic circumstances, lack of social support, recent life events and past history of depression have all been implicated. A striking risk factor, however, for older people is the disability associated with ill health. While poor health has also been found to be strongly associated with risk for depression in younger people, its importance in later life stems from the much higher prevalence of poor health and disability among older people. It has been estimated that up to 70% of all new cases of depression arising in older people may be caused by disability associated with ill health (Prince *et al*, 1998).

Most authors are agreed that, however and wherever it is assessed, late-life depression is a common and distressing disorder with important adverse consequences both for affected individuals and for society as a whole. The negative effects of depression on the functional capacity and well-being of older subjects are well established (Gurland, 1992). Unsurprisingly, older people with depression are relatively heavy users of a range of health and social care services (Livingston *et al*, 1990; Unutzer *et al*,1997). The costs of this excess service provision are likely to be considerable (Smith *et al*, 1995). Finally, late-life depression seems to confer an additional independent risk for death, even after adjusting for physical illness and disability (Brill *et al*, 1992; Abas *et al*, 2002).

Cognitive impairment

Cognitive function is the measured performance on any test of higher intellectual function, such as memory, attention/concentration, reasoning or mental speed. Cognitive impairment is defined as relatively poor function on one or more of these tests. In older age groups this category includes people with clinical dementia (cognitive function which has declined to a sufficient extent to impair activities of daily living), but will also include people with more mild cognitive decline and those whose cognitive function has been low throughout adult life. In any research study, there will also be some people with 'normal' cognitive function who for one reason or another score within the 'impaired' range for a test on the day it is administered. There is no absolute division between 'normal' and 'impaired' scores for cognitive tests. Any definition of cognitive impairment is based on a cut-off point for a particular test. Most of the data described in this report concern scores on the TICS-m test (described in more detail in section 1.4.2).

Relatively impaired cognitive function can, of course, be defined at any age. However it assumes particular importance for older age groups because of its association with underlying deterioration of brain function associated with the early stages of dementia, as well as with other health conditions which may cause impairment such as vascular disease and stroke. Cognitive impairment is strongly associated with increased age and lower educational attainment or duration. Associations with gender tend to vary according to the test which is administered (Huppert, 1987; Welsh *et al*, 1994).

Studies investigating cognitive impairment have tended to focus on dementia. However, more mild cognitive impairment potentially has a higher public health impact because of the much larger numbers of people affected. Level of cognitive function has important implications for older populations. Cognitive impairment has been found to be a strong risk factor for the later development of dementia (Elias *et al*, 2000), and is associated with an increased risk of other adverse outcomes such as mortality and stroke (Ferucci *et al*, 1996; Gale *et al*, 1996). It has also been found to be strongly associated with disability, and subsequent functional decline (Hébert *et al*, 1999; Zuccalà *et al*, 2001), as well as with depression (Fuhrer *et al*, 1992). Associations between cognitive impairment and increased service use would therefore be expected. A US study found that cognitive impairment strongly predicted hospitalisation over a one-year period, after adjustment for impaired activities of daily living (Weiler *et al*, 1991). Associations between cognitive impairment and longer duration of hospital stay have also been widely reported (Binder and Robins, 1990). However people with cognitive impairment have been found to be less likely to receive domiciliary services than people with equivalent physical disability (Ely *et al*, 1997), suggesting unmet need in community settings.

1.3 Coverage of the report

The main report from the 2000 survey of psychiatric morbidity among adults (PMA 2000) presented data on prevalence rates across the full range of psychiatric morbidity among people aged 16–74 living in private households (Singleton *et al*, 2001). This report focuses on survey participants aged 60–74. The prevalence of many mental health problems is low in this age group relative to that in other age groups so, in our sample, few individuals would have the rarer personality disorders, psychotic and severe affective disorders, and substance dependence. Impairment of cognitive function, on the other hand, is likely to be more prevalent with older age. For this reason, in the PMA 2000 survey, respondents aged 60 years and over completed a series of assessments of cognitive function. This report presents analysis relating to common mental disorders and cognitive function in people aged 60–74 years.

Chapter 2 presents the prevalence of common mental disorders, as indicated by overall levels of neurotic symptoms and for individual neurotic disorders, and the variation in prevalence associated with a range of social and economic characteristics:

- Marital status, household size, tenure, and area type.
- Educational attainment and social class.
- Income and receipt of benefits.

Chapter 3 examines the association between prevalence of neurotic symptoms and disorders and other factors – past and current experience of difficulties and stressful events, physical health problems and the level of social support available to them. In Chapter 4 the impact of common mental disorders on wellbeing and day-to-day functioning is explored. The indices of wellbeing and day-to-day functioning used are:

- Scores on the SF-12 (a self-report measure of mental and physical wellbeing).
- Difficulties with seven aspects of daily living.
- Use of health and community services.

Chapters 5 and 6 focus on variation in cognitive function, (cognitive impairment, crystallised intelligence, and verbal fluency) with the socio-demographic characteristics, and on the impact of cognitive function on day-to-day functioning, respectively. They parallel Chapters 2 and 4 on neurotic symptoms and disorders. Finally, Chapter 7 examines the association between neurotic symptoms and disorders, cognitive function, and physical illness.

1.4 Methods and concepts used in this report

1.4.1 Assessment and definitions of neurotic symptoms and neurotic disorders

This report covers prevalence of neurotic symptoms and disorders. A brief summary of how these disorders were assessed is given below. Further details can be found in Appendix B.

Neurotic symptoms and disorders in the week preceding interview were assessed using the revised version of the Clinical Interview Schedule, CIS-R (Lewis and Pelosi, 1990; Lewis *et al*, 1992). In Chapter 2 of this report, data are presented on distribution of total CIS-R scores, which give an indication of severity of symptoms, six individual neurotic disorders, and some groups of neurotic disorders.

The CIS-R is comprised of 14 sections each covering a particular type of neurotic symptom, as listed below.

- Somatic symptoms.
- Fatigue.
- Concentration and forgetfulness.
- Sleep problems.
- Irritability.
- Worry about physical health.
- Depression.
- Depressive ideas.
- Worry.
- Anxiety.
- Phobias.
- Panic.
- Compulsions.
- Obsessions.

Each section begins with a number of mandatory filter questions. They establish the existence of a particular neurotic symptom in the past month. A positive response leads to a more detailed assessment of the symptom in the past week: frequency, duration, severity, and time since onset. Answers to these questions determine the informant's score on each section. Possible scores range from zero to 4 on each section (except the section on depressive ideas, which has a maximum score of 5). The 14 individual symptom scores can be summed to give a total CIS-R score, which reflects the overall severity of neurotic symptoms. Total scores range between 0 and 57. A score of 12 or above indicates significant levels of neurotic symptoms, while a score of 18 or more suggests a level of symptoms likely to require treatment.

Specific neurotic disorders were assessed by looking at the answers to various sections of the CIS-R and applying algorithms based on the ICD-10 diagnostic criteria for research (World Health Organisation, 1993).

Six diagnostic categories can be obtained from the CIS-R:

- Generalised Anxiety Disorder.
- Depressive episode.
- Phobias.
- Obsessive Compulsive Disorder.
- Panic disorder.
- Mixed anxiety and depressive disorder.

An individual may concurrently have more than one category of neurotic disorder. All respondents with a CIS-R score of 12 or more will have a diagnosis of a neurotic disorder. Those with a score of 12 or above who cannot be classified otherwise are given the diagnosis of mixed anxiety and depressive disorder. However, there will also be respondents diagnosed with a neurotic disorder who do not have a CIS-R score of 12 or more. Further, while the specific diagnoses of neurotic disorders give no information about overall severity of the symptoms, CIS-R total scores do.

The following groupings of disorders are also used in this report:

- Any anxiety disorder (general anxiety disorder, panic, phobia and obsessive-compulsive disorder and mixed anxiety and depressive disorder).
- Any anxiety disorder excluding mixed anxiety and depressive disorder.
- Any neurotic disorder.

1.4.2 Assessment and definition of cognitive functioning

The three measures used were:

1. The TICS-m: a short measure to screen for dementia which includes a word list learning task. The version used in our battery comprised 13 items with a maximum score of 39 points (Plassman *et al*, 1995). The items included are shown in Appendix B. It tests orientation, concentration, immediate and delayed memory, naming, calculation, comprehension and reasoning. The test of immediate and delayed memory is a ten word list-learning exercise that can be scored separately. Three scores are therefore generated, the total TICS-m score, an immediate learning score, and a delayed learning score. The cut-off which was used to define 'cognitive impairment' in this survey is a TICS-m score below 21 (out of a possible 39). The TICS-m has been recently evaluated against the Mini-Mental State Examination (MMSE), another brief test in common clinical use, and the TICS-m 20|21 cut-off was found to be equivalent to a score of below 25 on the MMSE (de Jager *et al*, 2003).
2. The Animal Naming Test: a verbal fluency test from the Consortium to Establish a Registry for Alzheimer's Disease (CERAD) neurological test battery, which is scored as a count of the number of animals that the respondent can name in a minute.

3. The New Adult Reading Test (NART) (Nelson and Willison, 1991): this is a widely used and validated brief measure of crystallised (pre-morbid) intelligence for English speakers. Respondents are shown fifty words (the pronunciation of which cannot be readily guessed) and asked to read them. These are listed in Appendix B. The test is scored in terms of the number of words pronounced accurately. This NART score can be converted to an estimated verbal IQ score.

1.5 Sampling and interviewing procedures

The survey was carried out between March and September 2000. A two-stage approach to the assessment of mental disorders was used. The first stage interviews were carried out by ONS interviewers and included structured assessment and screening instruments for measuring mental disorders, as well as covering a range of other topics, such as service use, risk factors for disorder and background socio-demographic factors. A sub-sample of people were then selected to take part in a second stage interview to assess psychosis and personality disorder, the assessment of which requires a more detailed interview than was possible at the first stage and some clinical judgement. These additional interviews were carried out by specially trained psychologists employed by the University of Leicester. However, only data from the first stage interviews are included in this report.

The small users postcode address file (PAF) was used as the sampling frame for the survey, because of its good coverage of private households in Great Britain. In the PAF, the postcode sectors were stratified on the basis of socio-economic group within NHS Region. A postal sector contains on average 2,550 addresses.

Initially, 438 postal sectors (the primary sampling units) were selected with a probability proportional to size (number of addresses). This included 370 sectors in England, 22 in Wales and 46 in Scotland. Within each of these sectors, 36 delivery points were selected, yielding a sample of 15,804 addresses.

Interviewers visited the 15,804 addresses to identify private households with at least one person aged 16–74 years. The Kish grid method was then used to randomly select one person in each household. (Kish, 1965). The Psychiatric Morbidity among Adults Living in Private Households Technical Report gives details of the questionnaire and assessment, sampling and weighting procedures (Singleton *et al*, 2002) and is available on the National Statistics website (www.statistics.gov.uk).

Overall, 10% of sampled addresses were ineligible because they contained no private households. Of the remaining addresses, 11% contained no-one within the eligible age range, which left an eligible sample of 12,792 addresses.

In the whole survey sample, just under 70% of those approached agreed to take part in an interview and despite the length of the interview, 95% completed the full interview. Of the 8,580 people who participated in the survey, 1,989 were aged 60 or above and these make up the sample covered in this report.

1.6 Access to survey data

The anonymised data from the survey has been lodged with the Data Archive at the University of Essex. Independent researchers who wish to carry out their own analyses should apply to the Archive for access. For further information about archived data, please contact:

ESRC Data Archive
University of Essex
Wivenhoe Park
Colchester
Essex CO4 3SQ
Tel: (UK) 01206 872323
FAX: (UK) 01206 872003
Email: archive@essex.ac.uk.

References

Abas M, Prince M and Hotopf M. (2002) Depression and mortality in an at risk population: A follow-up of the MRC Elderly Hypertension study. *British Journal of Psychiatry* **181**, 123–128.

Beekman A T F, Copeland J R M and Prince M J (1999) Review of community prevalence of depression in later life. *British Journal of Psychiatry* **174**, 307–11.

Binder E F and Robins L N (1990). Cognitive impairment and length of hospital stay in older persons. *Journal of the American Geriatrics Society* **38**, 759–766.

Bland R C, Newman S C and Orn H (1988) Prevalence of psychiatric disorders in the elderly in Edmonton. *Acta Psychiatrica Scandinavica. Suppl.* **338**, 57–63.

Brill P A Kohl H W and Blair S N (1992) Anxiety, depression, physical fitness and all cause mortality in men. *Journal of Psychosomatic Research* **36**, 267–273.

Copeland J R M, Hooijer C, Jordan A, Lawlor B, Lobo A, Magnusson H, Mann A, Meller I, Prince M J, Reischies F M, Turrina C and Wilson K C M (1999) Depression in Europe: Geographical distribution among older people. *British Journal of Psychiatry* **174**, 312–21.

Coulthard M, Farrell M, Singleton N, Meltzer H, (2002) *Tobacco, alcohol and drug use and mental health*, TSO: London.

de Jager C A, Budge M.M and Clarke R (2003) Utility of the TICS-M for the assessment of cognitive function in older adults. *International Journal of Geriatric Psychiatry* **18**, 318–324.

Elias M F, Beiser A, Wolf P A, Au R, White R F and D'Agostino R B (2000) The preclinical phase of Alzheimer disease: a 22-year prospective study of the Framingham cohort. *Archives of Neurology* **57**, 808–813.

Ely M, Brayne C, Huppert F A, O'Connor D W and Pollitt P A (1997) Cognitive impairment: a challenge for community care. A comparison of the domiciliary service receipt of cognitively impaired and equally dependent physically impaired elderly women. *Age and Ageing* **26**, 301–308.

Ernst C and Angst J (1995) Depression in old age. Is there a real decrease in prevalence? A review. *European Archives of Psychiatry and Clinical Neuroscience* **245**, 272–287.

Ferucci L, Guralnik J M, Salive M E, Pahor M, Corti M, Baroni A and Havlik R J (1996) Cognitive impairment and risk of stroke in the older population. *Journal of the American Geriatrics Society* **44**, 237–241.

Fuhrer R, Antonucci T C, Gagnon M, Dartigues,J-F and Barberger-Gateau P (1992) Depressive symptomatology and cognitive functioning: an epidemiological survey in an elderly community sample in France. *Psychological Medicine* **22**, 159–172.

Gale, C. R., Martyn, C. N. and Cooper, C. (1996). Cognitive impairment and mortality in a cohort of elderly people. *British Medical Journal* **312**, 608-611.

Gurland B (1992) The impact of depression on quality of life of the elderly. *Clin Geriatr Med* **8(2)**, 377–86.

Hébert R, Brayne C and Spiegelhalter D (1999) Factors associated with functional decline and improvement in a very elderly community-dwelling population. *American Journal of Epidemiology* **150**, 501–510.

Henderson A S, Jorm A F, Mackinnon A, Christensen H *et al* (1993) The prevalence of depressive disorders and the distribution of depressive symptoms in later life: a survey using Draft ICD-10 and DSM-III-R. *Psychological Medicine* **23**, 719–729.

Huppert F A (1987) Cognitive function, in Cox B D, Blaxter M, Buckle A L J, Fenner N P, Golding J F, Gore M, Huppert F A, Nickson J, Roth,M, Stark J, Wadsworth M E J and Wichelow M J (eds.), *Health and Lifestyle Survey: preliminary report of a nationwide survey of the physical and mental health, attitudes and lifestyle of a random sample of 9003 British adults*, pp 43–50, Health Promotion Trust: London.

Kanowski S (1994) Age-dependent epidemiology of depression. *Gerontology* **40 Suppl 1**, 1–4.

Kish L (1965) *Survey Sampling*, Wiley & Sons: London:

Knauper B and Wittchen H U (1994) Diagnosing major depression in the elderly: evidence for response bias in standardized diagnostic interviews? *Journal of Psychiatric Research* **28**, 147–164.

Lewis G and Pelosi A J (1990) *Manual of the revised clinical interview schedule (CIS-R)*, Institute of Psychiatry: London.

Lewis G, Pelosi A, Araya R C and Dunn G (1992) Measuring psychiatric disorder in the community: a standardised assessment for use by lay interviewers. *Psychological Medicine* **22**, 465–486.

Livingston G, Thomas A, Graham N *et al* (1990) The Gospel Oak Project: the use of health and social services by dependent elderly people in the community. *Health Trends* **2**, 70–73.

Meltzer H, Lader D, Corbin C, Singleton N, Jenkins R and Brugha T (2002a) *Non-fatal suicidal behaviour among adults aged 16-74 in Great Britain*, TSO: London.

Meltzer H, Singleton N, Lee A, Bebbington P, Brugha T and Jenkins R (2002b) *The Social and Economic Circumstances of Adults with Mental Disorders*, TSO: London.

Nelson H E and Willison J R (1991) *National Adult Reading Test (NART) (Second Edition)*, NFER-Nelson: Windsor.

O'Brien M, Singleton N, Sparks J, Meltzer H and Brugha T (2002) *Adults with a psychotic disorder living in private households, 2000*, TSO: London.

Plassman B L, Welsh K A, Helms M, Brandt J, Page W F and Breitner J C (1995) Intelligence and education as predictors of cognitive state in late life: a 50-year follow-up. *Neurology* **45(8)**, 1446–50.

Prince M J, Harwood R, Thomas A and Mann A H (1998) A prospective population-based cohort study of the effects of disablement and social milieu on the onset and maintenance of late-life depression. Gospel Oak VII. *Psychological Medicine* **28**, 337–350.

Prince M J and Beekman A T F (2001) The aetiology of late-life depression, in John Copeland, M. Abou-Saleh and Dan Blazer (eds), *Principles and Practice of Geriatric Psychiatry* (second edition).

Singleton N, Bumpstead R, O'Brien M, Lee A and Meltzer H (2001) *Psychiatric morbidity among adults living in private households, 2000*, TSO: London.

Singleton N, Lee A and Meltzer H (2002) *Psychiatric morbidity among adults living in private households, 2000: Technical Report*, Office for National Statistics: London, (www.statistics.gov.uk).

Snowdon J (1990) The prevalence of depression in old age. *International Journal of Geriatric Psychiatry* **5**, 141–144.

Smith K, Shah A, Wright K *et al* (1995) The prevalence and costs of psychiatric disorders and learning disabilities. *British Journal of Psychiatry* **166,** 9–18.

Tannock C and Katona C (1995) Minor depression in the aged. Concepts, prevalence and optimal management. *Drugs and Ageing* **6**, 278–292.

Unutzer J, Patrick D L, Simon G *et al* (1997) Depressive Symptoms and the Cost of Health Services in HMO Patients Aged 65 Years and Older. *Journal of the American Medical Association* **277**, 1618–1623.

Weiler P G, Lubben J E and Chi I (1991). Cognitive impairment and hospital use. *American Journal of Public Health* **81**, 1153–1157.

Weissman M M, Leaf P J, Tischler G L, Blazer D G *et al* (1988) Affective disorders in five United States communities [published erratum appears in *Psychol Med* 1988 Aug;18(3):following 792]. *Psychological Medicine* **18**, 141–153.

Welsh K A, Butters N, Mohs R C, Beekly D, Edland S, Fillenbaum G and Heyman A (1994) The consortium to establish a registry for Alzheimer's disease. Part V. A normative study of the neuropsychological battery. *Neurology* **44**, 609–614.

World Health Organisation (1993) *The ICD-10 classification of mental and behavioural disorders: Diagnostic criteria for research*. World Health Organisation: Geneva.

Zuccalà G, Onder G, Pedone C, Cocchi A, Carosella L, Cattel C, Carbonin P U and Bernabei R (2001) Cognitive dysfunction as a major determinant of disability in patients with heart failure: results from a multicentre survey. *Journal of Neurology, Neurosurgery and Psychiatry* 70, 109–112.

2 Prevalence of common mental disorder

2.1 Introduction

This chapter reports the prevalence of common mental disorder (sometimes called neurotic disorders) and variations in prevalence between groups defined by a range of social and economic characteristics previously shown to be associated with psychiatric morbidity in adults. The characteristics examined were: age, marital status, household size, tenure, area type, educational qualifications, social class, income, sources of income, and region. Information on variations in prevalence of psychiatric morbidity associated with these variables in adults across the whole age range (16–74 years) can be found in the main report from the 2000 survey, *Psychiatric Morbidity Among Adults Living in Private Households, 2000*, (Singleton *et al*, 2001), and in a topic report from the 2000 survey, *The Social and Economic Circumstances of Adults with Mental Disorders* (Meltzer *et al*, 2002).

The tables accompanying this chapter present distributions of both total scores on the revised Clinical Interview Schedule (CIS-R) and of different types of neurotic disorders (see Chapter 1 and Appendix B for a description of the CIS-R and how this is used to provide assessments of neurotic disorders). For simplicity, since the associations for these two indices of mental health generally follow the same pattern, only the associations with CIS-R scores are reported in the text. The one exception to this is the reporting of variations in prevalence of neurotic disorder with age and sex.

Although data were collected about respondents' ethnicity in the survey these are not reported on here. The number of respondents, in the 60–74 year age group, not of 'White' ethnicity was very small, and very heterogeneous, which makes meaningful analysis of variation by ethnic group impossible.

In addition to the analyses of the associations between common mental disorder and individual socio-economic characteristics, this chapter reports on the results of a multivariate analysis, using logistic regression. This analysis, described in more detail in section 2.2.4, allows the evaluation of the association between individual socio-economic characteristics and CIS-R scores after taking account of the associations between CIS-R scores and other socio-economic characteristics.

2.2 Distribution of common mental disorder

2.2.1 *Personal and household characteristics*

One of the clearest and most reproducible findings in psychiatric epidemiology is the apparent excess of depression among women. It has been suggested that the extent of this excess varies across the life course, increasing from puberty into mid-life, and then declining gradually into late-life (Jorm, 1987). The EURODEP consortium (Prince *et al*, 1999) reported a clear cut excess of depression symptoms in older women in population-based studies from 13 out of 14 European centres. Interestingly, this association was consistently modified by marital status, with marriage being protective for men but a risk factor among women. It has also been previously reported that marriage is associated with relatively low mortality and good health, again with this protective effect being stronger for men than for women (Jacobs, 1977; Berkman and Syme, 1979).

In the entire 2000 adult survey sample (aged between 16 and 74), 15% of respondents had a CIS-R score of 12 or more and about 16% had a neurotic disorder (Singleton *et al*, 2001, pp 37 and 40). However older people, both men and women, generally showed a lower prevalence of disorder. Overall, one in ten of the sample of people aged 60–74 covered in this report (10%) had total CIS-R scores at or above the threshold of 12 which is taken to indicate the presence of common mental disorder. Half of these (5%) had scores of 18 or more, indicating a high level of symptoms likely to require treatment. About one in eight of all respondents (12%) had at least one of the six neurotic disorders covered in the CIS-R. The most prevalent disorder, 7%, was mixed anxiety and depressive disorder (the diagnosis applied to individuals who have high levels of neurotic symptoms but do not meet the criteria for any of the other disorders). The least prevalent was panic disorder, at less than 1%. *(Table 2.1 and 2.2)*

Women aged 60–74 years were more likely than men in that age group to have CIS-R scores at or above the threshold (12% of women compared with 8% of men), and to have any neurotic disorder (14% of women compared with 9% of men). They were also more likely to have the specific diagnoses of mixed anxiety and depressive disorder (8% of women compared with 5% of men), and generalised anxiety disorder (4% of women compared with 2% of men). This sex difference among the sample of older people parallels that observed in the entire 2000 adult survey sample, in which 18% of women compared with 12% of men had CIS-R scores of 12 or more (Singleton *et al*, 2001, pp 37 and 40). *(Table 2.1 and 2.2)*

In the whole adult survey sample (aged from 16 to 74) there were differences between men and women in the pattern of variation in prevalence of common mental disorder with age. Among men, both younger and older men had relatively low rates of disorder while the prevalence peaked in middle age. Thus, men were most likely to have common mental disorder (a CIS-R score of 12 or more) between the ages of 45 and 49 years (19%), and least likely to have common mental disorder between the ages of 16 and 24 (8%) and the ages of 65 to 74 (5%). Young women had similar levels of common mental disorder to those in the middle age bands and then the prevalence declined after about age 54. Women were most likely to have CIS-R scores of 12 or more between the ages of 50 and 54 years (23%), and least likely to have such scores at age 70 to 74 years (11%) (Singleton *et al*, 2001, p 37).

In the sample of older people covered in this report, respondents aged between 60 and 64 years reported a higher prevalence of neurotic symptoms than older respondents (13% of respondents aged 60–64 had CIS-R scores of 12 or more compared with 9% and 8% of those aged 65–69 and 70–74 years, respectively). They were also more likely to have a neurotic disorder, (15% of respondents aged 60–64 compared with 10% and 9% of those aged 65–69 and 70–74 years). *(Table 2.1 and 2.2)*

However, as was the case in the entire adult sample, these overall differences hide differences between men and women in the pattern of variation in the prevalence of common mental disorder with age. In particular, the age-associated decrease in prevalence was stronger in men than in women. Thirteen per cent of men aged 60–64 years, compared with just 5% of men aged 65–69 and 6% of men aged 70–74 years had CIS-R scores of 12 or more. Among women the proportions with CIS-R scores of 12 or more were 13% for those aged 60–64, 12% for those aged 65–69, and 11% for those aged 70–74 years. The variation in prevalence of neurotic disorders followed the same pattern. *(Table 2.1 and 2.2 and Figure 2.1)*

Figure 2.1 Proportion of people with a CIS-R score of 12 or above by age and sex

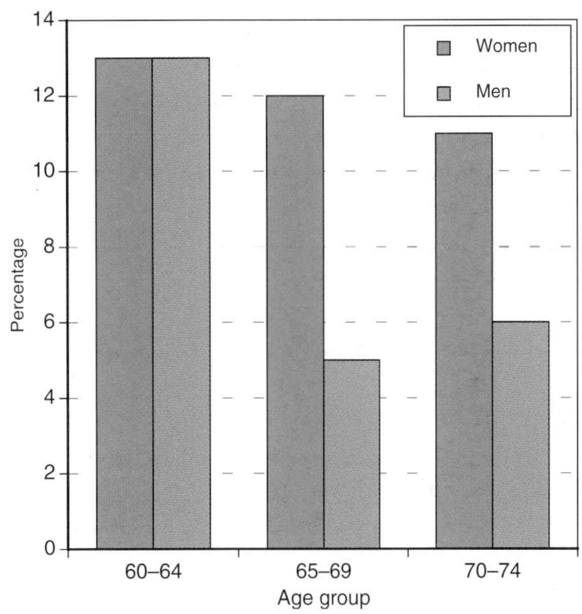

Consistent with previous studies described above, the pattern of variation in the prevalence of common mental disorder with marital status was different for men and women. Among both sexes, divorced and separated respondents were most likely to have CIS-R scores of 12 or more. Women who were single had a lower likelihood of mental health problems than any other women – just 4% of single women compared with 12% of married women, 12% of widowed women and 19% of divorced and separated women had CIS-R scores of 12 or more. Among men, the likelihood of having mental health problems was lowest in those who were married or single, and similar in these two groups. Seven per cent of married men, 8% of single men, 11% of widowed men, and 17% of divorced or separated men had CIS-R scores indicative of mental health problems.

(Tables 2.3 and 2.4)

The prevalence of common mental disorder varied significantly with household size only among men, and reflects the association between household size and marital status. Men living in two-person households were less likely than men living in single-person households to have CIS-R scores of 12 or more (6% compared with 14%). Although the distribution of CIS-R scores did not vary significantly with household size among women, it is noteworthy that women in two-person households were twice as likely as men in two-person households to have CIS-R scores of 12 or more (12% compared with 6%).

(Tables 2.3 and 2.4)

The likelihood of having mental health problems varied among groups with differing types of tenure in both sexes. Respondents who owned their homes outright were less likely than those living in rented accommodation to have CIS-R scores of 12 or more (7% compared with 18%). Among respondents owning their homes with a mortgage, the proportion with common mental disorder lay between the figures for the other two types of tenure (12% had a CIS-R score of 12 or more).

(Tables 2.5 and 2.6)

Area of residence was classified as urban, semi-rural or rural. Urban respondents in general were more likely than those in semi-rural and rural areas to have mental health problems. However, this difference was statistically significant only for the comparison between women living in urban and rural areas. Fourteen per cent of women living in urban areas compared with just 6% of women living in rural areas had CIS-R scores of 12 or more. *(Tables 2.5 and 2.6)*

2.2.2 Education and social class

Cross-sectional community surveys from a variety of cultures have reported associations between late-life depression and relative disadvantage in income (Berkman *et al*, 1986; Kennedy *et al* 1989; Murrell *et al*, 1983; Madianos *et al*, 1992; O'Hara *et al* 1985; Woo *et al*, 1994), housing status (Berkman *et al*, 1986; Murrell *et al*, 1983; Woo *et al*, 1994), and education (Berkman *et al*, 1986; Fuhrer *et al*, 1992; Kennedy *et al*, 1989; Murrell *et al*, 1983; Woo *et al*, 1994; O'Hara *et al*, 1985). These are, of course, highly correlated variables, and it will always be difficult to determine the effect of one, independent of the others. The possibility of reverse causality also needs to be considered because of the well-recognised phenomenon of social drift; people whose adult life has been scarred by depression may experience occupational and economic disadvantage.

In the current survey, although the likelihood of reporting significant levels of neurotic symptoms was lower among respondents with some educational qualifications than among those with no qualifications, the differences only reached statistical significance among women. Women with educational qualifications at 'A' level or above were less likely than women with no qualifications to have CIS-R scores of 12 or more (7% compared with 14%). *(Tables 2.7 and 2.8)*

The social class variable used in this report is based on information about an individual's last job, and is derived for all individuals who have ever been in paid employment. The result, in the sample of older people reported on here, is that the distribution of social class is very different between men and women. However, since the trends in the association between social class and mental health were similar in men and women only trends across the whole respondent sample are reported on. Men and women in manual social classes (III manual, IV, and V) were more likely than those in non-manual social classes (I or II combined, and III non-manual) to have a mental health problem. Likelihood of having a CIS-R score of 12 or more was highest among respondents in Social Class V and lowest among respondents in Social Classes I or II, (18% compared with 6% respectively had CIS-R scores above the threshold of 12).

(Tables 2.7 and 2.8)

2.2.3 Income

Socio-economic disadvantage has repeatedly been shown to be associated with health disadvantage. Analyses of data from the Health Survey for England for people aged 65–84 years showed a range of indicators of lower socio-economic status, and receipt of income support in particular, to be consistently associated with poorer outcomes on a range of indices of health, including scores on the

General Health Questionnaire, (a measure of minor psychiatric morbidity) (Grundy and Sloggett, 2003).

Consistent with these findings and those of other studies mentioned earlier, the likelihood of having a mental health problem, among the sample covered in this report, tended to decrease with increasing household income in both sexes. About one in seven respondents (15%) with a weekly household income below £200, compared with about one in ten (9%) of those with a weekly income of between £200 and £299 pounds, and just one in twenty-five people (4%) with a weekly income between £300 and £399 had CIS-R scores of 12 or more. Among respondents with a weekly household income of £500 or more, just one in fifty (2%) had a CIS-R score on or above the threshold of 12. *(Tables 2.9 and 2.10 and Figure 2.2)*

The age range of respondents described in this report is such that almost all women and around two-thirds of men would be in receipt of a state pension. Given the universality of this type of benefit, it was not surprising that receiving a state pension was not associated with increased likelihood of having a mental health problem. Similarly, since, for most respondents, the only benefit they received was a pension, receipt of benefits itself was not associated with increased likelihood of having mental health problem.

In contrast, being in receipt of the types of state benefit that are either means tested, or contingent upon disability, was associated with a greatly increased likelihood of having a mental health problem. At least one out of four respondents in receipt of any disability related benefit (27%), any income related benefit (29%), or housing benefit (24%) had CIS-R scores of 12 or more, making them at least twice as likely as respondents in general (10%) to have a CIS-R score above the threshold. *(Tables 2.11 and 2.12)*

As would be expected given the findings described above, respondents whose only source of regular income was state benefits were far more likely than other respondents to have mental health problems. About one in five respondents (19%) who relied on state benefits, compared with one in twenty respondents (5%) whose income came from sources other than state benefit only, and about one in twelve respondents (8%) whose income was a combination of state benefit and other sources, had CIS-R scores of 12 or more. *(Tables 2.13 and 2.14)*

Figure 2.2 Proportion of people with CIS-R scores of 12 or above by household income and sex

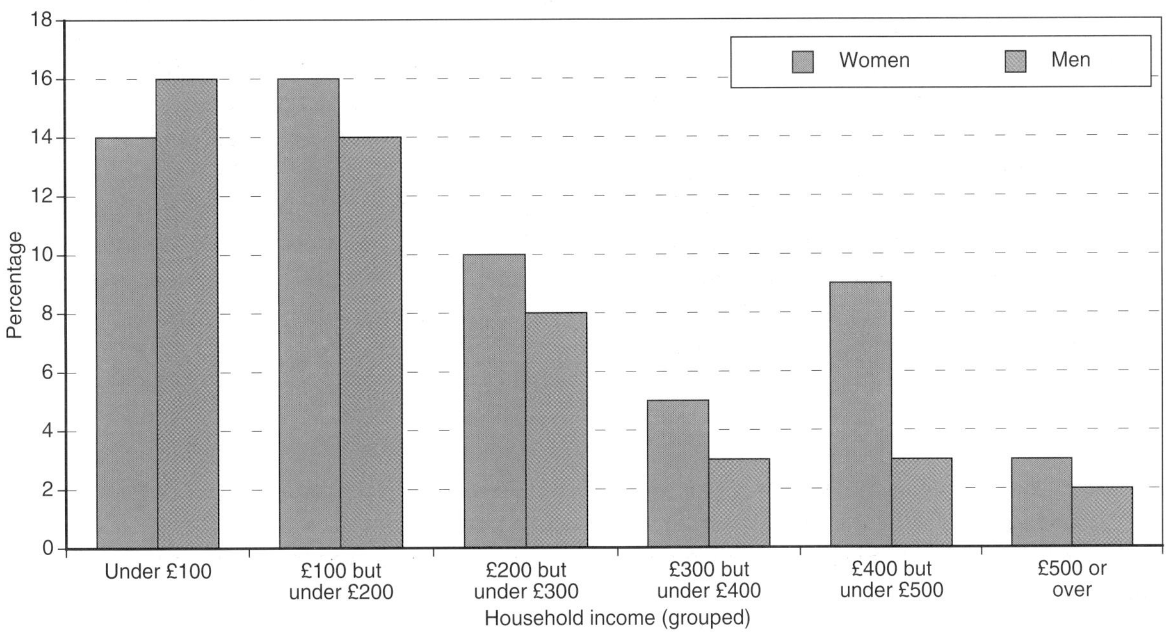

2.2.4 Independent associations between different socio-economic characteristics and CIS-R scores

The chapter so far has described the associations between a range of socio-economic characteristics and prevalence of common mental health problems, with each socio-economic characteristic considered individually. However, it is known that many socio-economic characteristics tend to be associated with each other. For example, people in higher social classes also tend to have higher income, and higher educational attainment. Similarly, older people are more likely than younger people to be widowed and retired. This means that an apparent association between one socio-economic variable and prevalence of mental health problems may in fact be due to the effect of another, closely associated, variable. One way of evaluating which of the variables examined in this report are most strongly, and independently, associated with mental health problems is to carry out logistic regression analyses.

Table 2.15 presents the results of the logistic regression looking at the association of a range of socio-economic characteristics with having a CIS-R score of 12 or above. The variables included in the model were: sex; age; marital status; household size; nature of tenure; type of area (urban, semi-rural, rural); level of highest educational qualifications; social class; household income; receipt of disability-related state benefits; receipt of income-related state benefits; receipt of housing benefit; receipt of any other type of state benefit. The characteristics examined in the analysis that were significantly associated with having a common mental disorder are set out in the first column of each table and the relative, independent, influence of each category of each characteristic is shown in the column headed 'Adjusted odds ratios'. The 'odds ratio' is the odds that respondents with a particular characteristic will have a CIS-R score of 12 or more divided by the odds of having a CIS-R score of 12 or more for respondents who do not have this characteristic. Characteristics that were not independently associated with CIS-R scores are listed at the bottom of the table.

In logistic regression analyses, for each characteristic one category is arbitrarily chosen as a standard against which the relative influence of other categories of the characteristic are compared. This reference category automatically has an odds ratio of 1. For example if sex is the characteristic of interest, and 'male' is set as the 'reference' category, then if the odds ratio for 'female' is 1.5 this means that women are more likely than men to show the behaviour of interest, in this case to have a common mental disorder. If the odds ratio for 'female' is 0.3, this means that women are less likely than men to show the behaviour of interest. For ease of interpretation, in this report, in most cases, the reference categories have been chosen so that they are the categories which were shown to be associated with least likelihood of having a CIS-R score of 12 or more in the analyses earlier in the chapter. (A further explanation of logistic regression analysis can be found in Appendix A).

The results in table 2.15 indicate that six of the socio-economic characteristics examined were independently associated with having clinically significant levels of neurotic symptoms. The six were: sex, social class, household income, receipt of a disability related state benefit, receipt of an income related state benefit, receipt of any other type of state benefit (this excludes housing benefit). Once other socio-economic characteristics were taken into account, the odds for women having high CIS-R scores were almost three times those for men (OR 2.90). Although social class was independently associated with CIS-R scores, only the odds for people in social class III manual were significantly different to those for people in the reference group, social classes I and II combined (OR 1.73). People in social class III manual were more likely than those in social classes I and II combined to have CIS-R scores of 12 or more.

After taking account of other socio-economic characteristics, people with a weekly household income below £300, and those whose income was not known, were significantly more likely than those with an income of £500 or more to have mental health problems. The odds for the income bands below £300 varied from 2.90 to 4.24, so they were around three or four times as high as those for the £500 or more income band. Although the odds ratios for people with incomes between £300 and £499 were greater than 1, the differences did not reach statistical significance. *(Table 2.15)*

Even after taking into account the effect of all the other income-related characteristics, people in receipt of a disability-related benefit, and people in receipt of an income-related benefit were significantly more likely than people not receiving these types of benefit to have mental health problems. The odds of having a CIS-R score of 12 or more for people receiving disability related benefit were more than four times those for people not receiving this type of benefit (OR 4.23). Being in receipt of income benefit more than doubled a person's odd of having a high CIS-R score (OR 2.39).

In contrast, people receiving any other forms of state benefit, (not disability-related, not income-related, and not housing benefit), were significantly less likely than those not receiving such benefits to have mental health problems. This initially surprising result may be accounted for by the fact that in most cases the 'other form of benefit' was a state retirement pension and that the majority of respondents would have been receiving this.
(Table 2.15)

These results support those of other research outlined above in indicating the importance financial disadvantage as a predictor of psychiatric morbidity in older age.

References

Berkman L F, Berkman C S, Kasl S, Freeman D H *et al* (1986) Depressive Symptoms in Relation to Physical Health and Functioning in the Elderly. *American Journal of Epidemiology* **124**, 372–388.

Berkman L F and Syme S L (1979) Social networks, host resistance, and mortality: a nine year follow-up study of Alameda county residents. *American Journal of Epidemiology* **109**, 186–204.

Fuhrer R, Antonucci T C, Gagnon M, Dartigues J-F and Barberger-Gateau P (1992) Depressive symptomatology and cognitive functioning: an epidemiological survey in an elderly community sample in France. *Psychological Medicine* **22**, 159–172.

Grundy E and Sloggett A (2003) Health inequalities in the older population: the role of personal capital, social resources and socio-economic circumstances, *Social Science and Medicine* 56, 935–947.

Jacobs S (1977) An epidemiological review of the mortality of bereavement. *Psychosomatic Medicine* **39**, 344–357.

Jorm A F (1987) Sex and age differences in depression: a quantitative synthesis of published research. *Aust N Z J Psychiatry* **21**, 46–53.

Kennedy G J, Kelman H R, Thomas C, Wisniewski W *et al* (1989) Hierarchy of Characteristics Associated With Depressive Symptoms in an Urban Elderly Sample. *American Journal of Psychiatry* **146**, 220–225.

Madianos M G, Gournas G and Stefanis C N (1992) Depressive symptoms and depression among elderly people in Athens. *Acta Psychiatrica Scandinavica.* **86**, 320–26.

Meltzer H, Singleton N, Lee A, Bebbington P, Brugha T and Jenkins R (2002) *The Social and Economic Circumstances of Adults with Mental Disorders*, TSO: London.

Murrell S A, Himmelfarb S and Wright K (1983) Prevalence of depression and its correlates in older adults. *American Journal of Epidemiology* **117**, 173–185.

O'Hara M W, Kohout F J and Wallace R B (1985) Depression among the rural elderly. A study of prevalence and correlates. *Journal of Nervous and Mental Disorders* **173**, 582–589.

Prince M J, Beekman A T F, Deeg D J H, Fuhrer R *et al* (1999) Depression symptoms in late life assessed using the EURO-D scale. *British Journal of Psychiatry* **174**, 339–345.

Singleton N, Bumpstead R, O'Brien M, Lee A and Meltzer H (2001) *Psychiatric morbidity among adults living in private households, 2000*, TSO: London.

Woo J, Ho S C, Lau J, Yuen Y K *et al* (1994) The prevalence of depressive symptoms and predisposing factors in an elderly Chinese population. *Acta Psychiatrica Scandinavica* **89**, 8–13.

Table 2.1 Distribution of CIS-R Score (grouped) by age and sex

	Age			
	60–64	65–69	70–74	All
	%	%	%	%
CIS-R score				
Women				
0–5	71	73	72	72
6–11	16	15	17	16
Under 12	**87**	**88**	**89**	**88**
12–17	7	7	7	7
18 and over	6	6	4	5
12 and over	**13**	**12**	**11**	**12**
Base	*407*	*373*	*356*	*1136*
Men				
0–5	74	84	85	80
6–11	13	12	9	12
Under 12	**87**	**95**	**94**	**92**
12–17	6	3	4	4
18 and over	7	2	2	4
12 and over	**13**	**5**	**6**	**8**
Base	*332*	*295*	*244*	*871*
All adults				
0–5	73	78	78	76
6–11	14	13	14	14
Under 12	**87**	**91**	**92**	**90**
12–17	6	5	5	6
18 and over	7	4	3	5
12 and over	**13**	**9**	**8**	**10**
Base	*739*	*668*	*600*	*2007*

Table 2.2 Prevalence of neurotic disorders by age and sex

	Age			
	60–64	65–69	70–74	All
	Percentage with each disorder			
Neurotic disorder				
Women				
Mixed anxiety and depressive disorder	9	8	7	8
Generalised anxiety disorder	4	4	3	4
Depressive episode	1	1	2	1
All phobias	2	1	0	1
Obsessive compulsive disorder	1	0	0	1
Panic disorder	-	1	1	0
Any anxiety disorder	5	6	4	5
Any neurotic disorder excluding mixed anxiety and depressive disorder	6	6	5	6
Any neurotic disorder	**15**	**15**	**12**	**14**
Base	*407*	*373*	*356*	*1136*
Men				
Mixed anxiety and depressive disorder	7	3	4	5
Generalised anxiety disorder	4	1	2	2
Depressive episode	3	0	1	2
All phobias	1	-	0	1
Obsessive compulsive disorder	1	-	-	0
Panic disorder	0	-	-	0
Any anxiety disorder	5	1	2	3
Any neurotic disorder excluding mixed anxiety and depressive disorder	7	2	3	4
Any neurotic disorder	**15**	**5**	**7**	**9**
Base	*332*	*295*	*244*	*871*
All Adults				
Mixed anxiety and depressive disorder	8	6	6	7
Generalised anxiety disorder	4	3	2	3
Depressive episode	2	1	1	1
All phobias	1	1	0	1
Obsessive compulsive disorder	1	0	0	1
Panic disorder	0	0	0	0
Any anxiety disorder	5	4	3	4
Any neurotic disorder excluding mixed anxiety and depressive disorder	7	4	4	5
Any neurotic disorder	**15**	**10**	**9**	**12**
Base	*739*	*668*	*600*	*2007*

Table 2.3 Distribution of CIS-R Score (grouped) by marital status, household size and sex

	Marital status				Household Size			
	Married	Divorced/ Separated	Single	Widowed	1	2	3 and over	All
	%	%	%	%	%	%	%	%
CIS-R score								
Women								
0–5	73	61	88	71	69	74	69	72
6–11	15	20	8	16	17	14	19	16
Under 12	**88**	**81**	**96**	**88**	**87**	**88**	**88**	**88**
12–17	7	10	4	7	8	6	9	7
18 and over	5	9	-	6	6	6	3	5
12 and over	**12**	**19**	**4**	**12**	**13**	**12**	**12**	**12**
Base	*591*	*133*	*67*	*345*	*479*	*581*	*76*	*1136*
Men								
0–5	81	67	84	82	75	83	71	80
6–11	12	16	8	7	10	11	16	12
Under 12	**93**	**83**	**92**	**89**	**86**	**94**	**87**	**92**
12-17	4	6	4	8	8	3	5	4
18 and over	3	11	4	3	6	3	7	4
12 and over	**7**	**17**	**8**	**11**	**14**	**6**	**13**	**8**
Base	*580*	*104*	*72*	*115*	*236*	*561*	*74*	*871*
All adults								
0–5	77	63	86	74	71	79	70	76
6–11	14	18	8	14	15	13	17	14
Under 12	**91**	**82**	**94**	**88**	**86**	**91**	**88**	**90**
12–17	5	8	4	7	8	5	7	6
18 and over	4	10	2	5	6	4	5	5
12 and over	**9**	**18**	**6**	**12**	**14**	**9**	**12**	**10**
Base	*1171*	*237*	*139*	*460*	*715*	*1142*	*150*	*2007*

Table 2.4 Prevalence of neurotic disorders

by marital status, household size and sex

	Marital status				Household Size			
	Married	Divorced/ Separated	Single	Widowed	1	2	3 and over	All
	Percentage with each disorder							
Neurotic disorder								
Women								
Mixed anxiety and depressive disorder	9	9	4	6	7	8	12	8
Generalised anxiety disorder	4	8	-	3	4	4	2	4
Depressive episode	1	4	-	2	3	1	-	1
All phobias	1	3	-	2	2	1	-	1
Obsessive compulsive disorder	0	2	-	1	1	1	-	1
Panic disorder	0	-	-	1	1	0	-	0
Any anxiety disorder	5	9	-	6	6	5	2	5
Any neurotic disorder excluding mixed anxiety and depressive disorder	5	10	-	8	7	6	2	6
Any neurotic disorder	**14**	**19**	**4**	**14**	**15**	**14**	**14**	**14**
Base	*591*	*133*	*67*	*345*	*479*	*581*	*76*	*1136*
Men								
Mixed anxiety and depressive disorder	5	6	4	9	8	4	8	5
Generalised anxiety disorder	2	8	2	2	4	2	1	2
Depressive episode	1	8	1	-	3	1	5	2
All phobias	1	1	1	-	1	0	2	1
Obsessive compulsive disorder	0	2	-	1	1	0	-	0
Panic disorder	0	-	1	-	0	0	-	0
Any anxiety disorder	2	9	5	2	5	2	3	3
Any neurotic disorder excluding mixed anxiety and depressive disorder	3	12	5	2	7	3	6	4
Any neurotic disorder	**8**	**18**	**8**	**11**	**15**	**7**	**14**	**9**
Base	*580*	*104*	*72*	*115*	*236*	*561*	*74*	*871*
All Adults								
Mixed anxiety and depressive disorder	7	7	4	7	8	6	10	7
Generalised anxiety disorder	3	8	1	3	4	3	2	3
Depressive episode	1	6	1	2	3	1	2	1
All phobias	1	2	1	2	1	1	1	1
Obsessive compulsive disorder	0	2	-	1	1	1	-	1
Panic disorder	0	-	1	1	1	0	-	0
Any anxiety disorder	4	9	2	5	6	4	3	4
Any neurotic disorder excluding mixed anxiety and depressive disorder	4	11	2	6	7	5	4	5
Any neurotic disorder	**11**	**19**	**7**	**13**	**15**	**10**	**14**	**12**
Base	*1171*	*237*	*139*	*460*	*715*	*1142*	*150*	*2007*

Table 2.5 Distribution of CIS-R Scores (grouped)

by tenure and area type and sex

	Tenure			Area type			
	Owned Outright	Owned with mortgage	Rented	Urban	Semi-rural	Rural	All
	%	%	%	%	%	%	%
CIS-R score							
Women							
0-5	75	67	66	70	75	79	72
6-11	15	20	15	16	15	15	16
Under 12	**91**	**87**	**81**	**86**	**90**	**94**	**88**
12-17	6	5	10	7	7	4	7
18 and over	4	7	9	7	3	1	5
12 and over	**9**	**13**	**19**	**14**	**10**	**6**	**12**
Base	*717*	*101*	304	726	306	104	1136
Men							
0-5	84	77	71	78	81	87	80
6-11	11	12	13	12	13	6	12
Under 12	**95**	**89**	**84**	**90**	**94**	**93**	**92**
12-17	3	5	9	5	2	5	4
18 and over	2	7	7	4	4	2	4
12 and over	**5**	**11**	**16**	**10**	**6**	**7**	**8**
Base	*547*	*129*	*181*	*498*	*246*	*127*	*871*
All adults							
0-5	79	73	68	74	78	84	76
6-11	13	15	14	15	14	10	14
Under 12	**93**	**88**	**82**	**88**	**92**	**94**	**90**
12-17	5	5	10	6	5	5	6
18 and over	3	7	9	6	4	2	5
12 and over	**7**	**12**	**18**	**12**	**8**	**6**	**10**
Base	*1264*	*230*	*485*	*1224*	*552*	*231*	*2007*

Table 2.6 Prevalence of neurotic disorders

by tenure and area type and sex

	Tenure			Area type			
	Owned Outright	Owned with mortgage	Rented	Urban	Semi-rural	Rural	All
	Percentage with each disorder						
Neurotic disorder							
Women							
Mixed anxiety and depressive disorder	6	7	13	9	8	3	8
Generalised anxiety disorder	3	7	4	5	2	3	4
Depressive episode	1	2	3	2	0	-	1
All phobias	1	-	2	2	1	1	1
Obsessive compulsive disorder	0	1	2	1	1	-	1
Panic disorder	1	-	0	0	1	-	0
Any anxiety disorder	4	7	6	6	4	3	5
Any neurotic disorder excluding mixed anxiety and depressive disorder	5	9	7	7	4	3	6
Any neurotic disorder	**11**	**16**	**21**	**16**	**13**	**6**	**14**
Base	*717*	*101*	*304*	*726*	*306*	*104*	*1136*
Men							
Mixed anxiety and depressive disorder	4	6	10	5	4	6	5
Generalised anxiety disorder	2	3	4	3	2	1	2
Depressive episode	0	4	3	2	1	-	2
All phobias	-	0	2	1	-	-	1
Obsessive compulsive disorder	0	0	1	1	0	-	0
Panic disorder	-	-	1	0	-	-	0
Any anxiety disorder	2	4	6	4	2	1	3
Any neurotic disorder excluding mixed anxiety and depressive disorder	2	6	9	5	3	1	4
Any neurotic disorder	**6**	**12**	**18**	**11**	**7**	**7**	**9**
Base	*547*	*129*	*181*	*498*	*246*	*127*	*871*
All Adults							
Mixed anxiety and depressive disorder	5	6	12	7	6	5	7
Generalised anxiety disorder	2	5	4	4	2	2	3
Depressive episode	1	3	3	2	1	-	1
All phobias	0	0	2	1	0	0	1
Obsessive compulsive disorder	0	1	2	1	1	-	1
Panic disorder	0	-	1	0	1	-	0
Any anxiety disorder	3	5	6	5	3	2	4
Any neurotic disorder excluding mixed anxiety and depressive disorder	4	7	8	6	4	2	5
Any neurotic disorder	**9**	**14**	**20**	**13**	**10**	**7**	**12**
Base	*1264*	*230*	*485*	*1224*	*552*	*231*	*2007*

Table 2.7 Distribution of CIS-R Score (grouped)

by highest educational qualification, social class and sex

	Highest educational qualifications			Social class					
	A level or above	GCSE level	No qualifications	I and II	III Non-manual	III Manual	IV	V	All
	%	%	%	%	%	%	%	%	%
CIS–R score									
Women									
0–5	78	76	69	75	74	69	69	64	72
6–11	15	13	17	15	18	14	16	16	16
Under 12	**93**	**88**	**86**	**90**	**91**	**83**	**85**	**80**	**88**
12–17	4	7	8	5	6	11	8	8	7
18 and over	3	5	6	5	3	6	7	12	5
12 and over	**7**	**12**	**14**	**10**	**9**	**17**	**15**	**20**	**12**
Base	*173*	*289*	*663*	*263*	*402*	*108*	*202*	*126*	*1136*
Men									
0–5	84	84	76	86	87	71	85	78	80
6–11	9	10	14	10	10	15	7	9	12
Under 12	**93**	**94**	**90**	**96**	**97**	**87**	**92**	**86**	**92**
12–17	3	2	6	3	1	6	4	11	4
18 and over	4	4	4	1	2	7	4	3	4
12 and over	**7**	**6**	**10**	**4**	**3**	**13**	**8**	**14**	**8**
Base	*213*	*187*	*456*	*309*	*68*	*304*	*121*	*46*	*871*
All adults									
0–5	82	79	72	81	76	71	75	68	76
6–11	12	12	16	12	16	15	12	14	14
Under 12	**93**	**91**	**88**	**94**	**92**	**86**	**88**	**82**	**90**
12–17	3	5	7	4	5	8	6	9	6
18 and over	4	4	5	3	3	7	6	9	5
12 and over	**7**	**9**	**12**	**6**	**8**	**14**	**12**	**18**	**10**
Base	*386*	*476*	*1119*	*572*	*470*	*412*	*323*	*172*	*2007*

Table 2.8 Prevalence of neurotic disorders

by highest educational qualification, social class and sex

	Highest educational qualifications			Social class					
	A level or above	GCSE level	No qualifications	I and II	III Non-manual	III Manual	IV	V	All
	Percentage with each disorder								
Neurotic disorder									
Women									
Mixed anxiety and depressive disorder	4	9	9	6	6	15	10	12	8
Generalised anxiety disorder	4	3	4	5	2	2	4	5	4
Depressive episode	0	1	2	0	1	2	2	4	1
All phobias	1	0	2	-	-	-	2	3	1
Obsessive compulsive disorder	-	0	1	-	1	1	1	1	1
Panic disorder	-	-	1	-	1	1	1	-	0
Any anxiety disorder	5	3	6	5	3	3	6	7	5
Any neurotic disorder excluding mixed anxiety and depressive disorder	5	3	8	5	5	4	7	10	6
Any neurotic disorder	**9**	**12**	**16**	**11**	**10**	**18**	**17**	**22**	**14**
Base	*173*	*289*	*663*	*263*	*402*	*108*	*202*	*126*	*1136*
Men									
Mixed anxiety and depressive disorder	4	3	6	3	-	8	3	14	5
Generalised anxiety disorder	2	2	3	1	1	3	5	1	2
Depressive episode	1	2	2	0	2	3	2	-	2
All phobias	0	-	1	-	-	-	-	-	
Obsessive compulsive disorder	1	-	1	-	-	1	-	-	0
Panic disorder	-	-	0	-	-	0	1	-	0
Any anxiety disorder	2	2	4	2	2	4	6	1	3
Any neurotic disorder excluding mixed anxiety and depressive disorder	3	4	5	2	3	6	6	1	4
Any neurotic disorder	**7**	**7**	**11**	**5**	**3**	**14**	**10**	**14**	**9**
Base	*213*	*187*	*456*	*309*	*68*	*304*	*121*	*46*	*871*
All Adults									
Mixed anxiety and depressive disorder	4	6	8	4	5	10	8	13	7
Generalised anxiety disorder	3	2	4	3	2	3	5	4	3
Depressive episode	1	1	2	0	1	3	2	3	1
All phobias	1	0	1	0	1	1	1	2	1
Obsessive compulsive disorder	0	0	1	-	1	1	0	1	1
Panic disorder	-	-	1	-	1	0	1	-	0
Any anxiety disorder	3	2	5	3	3	4	6	6	4
Any neurotic disorder excluding mixed anxiety and depressive disorder	4	3	6	4	4	6	7	7	5
Any neurotic disorder	**8**	**10**	**14**	**8**	**9**	**15**	**14**	**20**	**12**
Base	*386*	*476*	*1119*	*572*	*470*	*412*	*323*	*172*	*2007*

Table 2.9 Distribution of CIS-R Score (grouped) by household income (grouped) and sex

	Household weekly gross income						
	Under £100	£100 but under £200	£200 but under £300	£300 but under £400	£400 but under £500	£500 or over	All
	%	%	%	%	%	%	%
CIS–R score							
Women							
0–5	71	66	74	75	83	85	72
6–11	15	18	16	20	7	12	16
Under 12	**86**	**84**	**90**	**95**	**91**	**97**	**88**
12–17	7	9	5	4	7	2	7
18 and over	7	7	6	2	2	2	5
12 and over	**14**	**16**	**10**	**5**	**9**	**3**	**12**
Base	*179*	*401*	*194*	*93*	*43*	*62*	*1136*
Men							
0–5	75	72	83	79	88	85	80
6–11	9	14	8	19	9	13	12
Under 12	**84**	**86**	**92**	**97**	**97**	**98**	**92**
12–17	7	8	3	2	3	2	4
18 and over	9	6	5	1	-	-	4
12 and over	**16**	**14**	**8**	**3**	**3**	**2**	**8**
Base	*79*	*223*	*185*	*107*	*64*	*103*	*871*
All adults							
0–5	72	69	79	77	86	85	76
6–11	13	16	12	19	8	13	14
Under 12	**85**	**85**	**91**	**96**	**94**	**98**	**90**
12–17	7	9	4	3	5	2	6
18 and over	7	6	5	1	1	1	5
12 and over	**15**	**15**	**9**	**4**	**6**	**2**	**10**
Base	*258*	*624*	*379*	*200*	*107*	*165*	*2007*

Table 2.10 Prevalence of neurotic disorders by household income (grouped) and sex

	Household weekly gross income						
	Under £100	£100 but under £200	£200 but under £300	£300 but under £400	£400 but under £500	£500 or over	All
	Percentage with each disorder						
Neurotic disorder							
Women							
Mixed anxiety and depressive disorder	8	12	7	3	7	2	8
Generalised anxiety disorder	3	4	3	4	2	4	4
Depressive episode	3	1	1	-	-	-	1
All phobias	3	1	1	1	-	-	1
Obsessive compulsive disorder	1	1	1	-	-	-	1
Panic disorder	1	1	-	1	-	-	0
Any anxiety disorder	6	5	4	6	2	4	5
Any neurotic disorder excluding mixed anxiety and depressive disorder	8	6	5	6	2	4	6
Any neurotic disorder	**16**	**18**	**12**	**9**	**9**	**6**	**14**
Base	*179*	*401*	*194*	*93*	*43*	*62*	*1136*
Men							
Mixed anxiety and depressive disorder	14	8	5	2	3	2	5
Generalised anxiety disorder	5	5	2	2	-	-	2
Depressive episode	1	2	2	-	-	-	2
All phobias	-	1	-	-	-	2	1
Obsessive compulsive disorder	-	1	0	-	-	-	0
Panic disorder	-	1	-	-	-	-	0
Any anxiety disorder	5	6	2	2	-	2	3
Any neurotic disorder excluding mixed anxiety and depressive disorder	5	7	4	2	-	2	4
Any neurotic disorder	**19**	**15**	**9**	**4**	**3**	**4**	**9**
Base	*79*	*223*	*185*	*107*	*64*	*103*	*871*
All Adults							
Mixed anxiety and depressive disorder	10	10	6	2	5	2	7
Generalised anxiety disorder	4	4	3	3	1	1	3
Depressive episode	2	1	2	-	-	-	1
All phobias	2	1	1	0	-	1	1
Obsessive compulsive disorder	1	1	1	-	-	-	1
Panic disorder	1	1	-	1	-	-	0
Any anxiety disorder	6	5	3	4	1	3	4
Any neurotic disorder excluding mixed anxiety and depressive disorder	7	6	5	4	1	3	5
Any neurotic disorder	**17**	**17**	**10**	**6**	**6**	**4**	**12**
Base	*258*	*624*	*379*	*200*	*107*	*165*	*2007*

Table 2.11 Distribution of CIS-R Score (grouped)

by type of state benefits received and sex

	State Benefits Received						
	Any disability related	Any income related	Any other (eg Pensions, or child related)	Household receives housing benefit	Any benefits	No benefits	All
	%	%	%	%	%	%	%
CIS–R score							
Women							
0–5	46	49	72	60	72	69	72
6–11	23	19	16	14	16	11	16
Under 12	**69**	**68**	**88**	**74**	**88**	**80**	**88**
12–17	18	16	6	13	7	12	7
18 and over	13	16	5	12	5	8	5
12 and over	**31**	**32**	**12**	**26**	**12**	**20**	**12**
Base	*149*	*91*	*1067*	*186*	*1086*	*37*	*1136*
Men							
0–5	51	59	83	61	77	89	80
6–11	24	14	11	17	13	6	12
Under 12	**75**	**73**	**94**	**78**	**91**	**95**	**92**
12–17	11	10	4	11	4	5	4
18 and over	14	17	2	11	5	-	4
12 and over	**25**	**27**	**6**	**22**	**9**	**5**	**8**
Base	*185*	*69*	*554*	*108*	*667*	*187*	*871*
All adults							
0–5	49	54	76	61	74	86	76
6–11	24	17	14	15	15	7	14
Under 12	**73**	**71**	**91**	**76**	**89**	**93**	**90**
12–17	14	13	5	12	6	6	6
18 and over	13	16	4	12	5	1	5
12 and over	**27**	**29**	**9**	**24**	**11**	**7**	**10**
Base	*334*	*160*	*1621*	*294*	*1753*	*224*	*2007*

Table 2.12 Prevalence of neurotic disorders

by type of state benefits received and sex

	State Benefits Received						
	Any disability related	Any income related	Any other (eg Pensions, or child related)	Household receives housing benefit	Any benefits	No benefits	All
	Percentage with each disorder						
Neurotic disorder							
Women							
Mixed anxiety and depressive disorder	24	21	7	18	8	12	8
Generalised anxiety disorder	7	6	3	4	3	14	4
Depressive episode	3	7	1	4	1	3	1
All phobias	1	-	1	2	1	-	1
Obsessive compulsive disorder	1	2	1	2	1	-	1
Panic disorder	1	3	0	1	0	-	0
Any anxiety disorder	8	9	5	7	5	14	5
Any neurotic disorder excluding mixed anxiety and depressive disorder	9	11	6	8	6	14	6
Any neurotic disorder	**32**	**32**	**13**	**26**	**13**	**26**	**14**
Base	*149*	*91*	*1067*	*186*	*1086*	*37*	*1136*
Men							
Mixed anxiety and depressive disorder	13	17	4	15	6	4	5
Generalised anxiety disorder	6	9	1	5	3	2	2
Depressive episode	7	5	1	3	2	-	2
All phobias	2	3	0	1	1	-	1
Obsessive compulsive disorder	2	2	-	1	1	-	0
Panic disorder	1	1	-	2	0	-	0
Any anxiety disorder	9	11	2	7	3	2	3
Any neurotic disorder excluding mixed anxiety and depressive disorder	13	12	2	9	5	2	4
Any neurotic disorder	**27**	**29**	**7**	**24**	**10**	**5**	**9**
Base	*185*	*69*	*554*	*108*	*667*	*187*	*871*
All Adults							
Mixed anxiety and depressive disorder	17	19	6	17	7	5	7
Generalised anxiety disorder	7	8	3	5	3	4	3
Depressive episode	6	6	1	3	2	0	1
All phobias	2	1	1	2	1	-	1
Obsessive compulsive disorder	2	2	1	2	1	-	1
Panic disorder	1	2	0	1	0	-	0
Any anxiety disorder	8	10	4	7	4	4	4
Any neurotic disorder excluding mixed anxiety and depressive disorder	11	12	4	9	5	4	5
Any neurotic disorder	**29**	**31**	**11**	**25**	**12**	**9**	**12**
Base	*334*	*160*	*1621*	*294*	*1753*	*224*	*2007*

Table 2.13 Distribution of CIS-R Score (grouped) by sources of income and sex

	Sources of Regular Income			
	State benefits only	Non-state benefit income only	State benefits and other income	All
	%	%	%	%
CIS-R score				
Women				
0–5	68	77	74	72
6–11	15	14	16	16
Under 12	**83**	**90**	**90**	**88**
12–17	10	7	5	7
18 and over	7	3	4	5
12 and over	**17**	**10**	**10**	**12**
Base	*297*	*31*	*784*	*1136*
Men				
0–5	59	89	81	80
6–11	16	6	13	12
Under 12	**75**	**95**	**94**	**92**
12–17	13	5	3	4
18 and over	12	-	3	4
12 and over	**25**	**5**	**6**	**8**
Base	*121*	*184*	*544*	*871*
All adults				
0–5	65	87	77	76
6–11	15	7	15	14
Under 12	**81**	**95**	**92**	**90**
12–17	11	5	4	6
18 and over	9	0	4	5
12 and over	**19**	**5**	**8**	**10**
Base	*418*	*215*	*1328*	*2007*

Table 2.14 Prevalence of neurotic disorders by sources of income and sex

	Sources of Regular Income			
	State benefits only	Non-state benefit income only	State benefit and other income	All
	Percentage with each disorder			
Neurotic disorder				
Women				
Mixed anxiety and depressive disorder	11	7	7	8
Generalised anxiety disorder	4	11	3	4
Depressive episode	2	-	1	1
All phobias	1	-	1	1
Obsessive compulsive disorder	1	-	1	1
Panic disorder	1	-	0	0
Any anxiety disorder	6	11	4	5
Any neurotic disorder excluding mixed anxiety and depressive disorder	8	11	5	6
Any neurotic disorder	**19**	**18**	**11**	**14**
Base	*297*	*31*	*784*	*1136*
Men				
Mixed anxiety and depressive disorder	16	4	4	5
Generalised anxiety disorder	5	2	2	2
Depressive episode	7	-	1	2
All phobias	1	-	1	1
Obsessive compulsive disorder	-	-	1	0
Panic disorder	1	-	-	0
Any anxiety disorder	7	2	3	3
Any neurotic disorder excluding mixed anxiety and depressive disorder	11	2	4	4
Any neurotic disorder	**26**	**5**	**7**	**9**
Base	*121*	*184*	*544*	*871*
All Adults				
Mixed anxiety and depressive disorder	13	4	5	7
Generalised anxiety disorder	4	3	3	3
Depressive episode	3	-	1	1
All phobias	1	-	1	1
Obsessive compulsive disorder	1	-	1	1
Panic disorder	1	-	0	0
Any anxiety disorder	6	3	3	4
Any neurotic disorder excluding mixed anxiety and depressive disorder	9	3	4	5
Any neurotic disorder	**21**	**7**	**9**	**12**
Base	*418*	*215*	*1328*	*2007*

Table 2.15 Odds ratios for socio-economic characteristics associated with CIS-R scores of 12 and over

	Adjusted Odds Ratios	(95% confidence interval)
Socio–economic characteristics		
Sex		
Male	1.00	———
Female	2.90 **	(1.93–4.39)
Social Class		
I or II	1.00	———
III non manual	0.74	(0.44–1.25)
III manual	1.73 *	(1.07–2.79)
IV	1.05	(0.62–1.75)
V	1.56	(0.89–2.75)
Not known	0.25	(0.03–1.85)
Household Income		
Not known	3.96 *	(1.36–11.53)
Less than £100	3.71 *	(1.22–11.29)
£100 to £199	4.24 **	(1.49–12.08)
£200 to £299	2.90 *	(0.99–8.44)
£300 to £399	1.27	(0.38–4.29)
£400 to £499	2.27	(0.64–7.98)
£500 and over	1.00	———
Receives a disability related state benefit?		
No	1.00	———
Yes	4.23 **	(3.01–5.94)
Receives an income related state benefit?		
No	1.00	———
Yes	2.39 **	(1.49–3.83)
Receives any other type of state benefit? (Excludes housing benefit)		
No	1.00	———
Yes	0.49 **	(0.32–0.73)

Other variables not significantly associated:
Age, marital status, tenure in accommodation, area type, educational qualifications, receipt of housing benefit

* = $p < 0.05$
** = $p < 0.01$

3 Social context and physical health factors associated with common mental disorder

3.1 Introduction

Chapter 2 described the prevalence of common mental disorder associated with a range of demographic and socio-economic characteristics. This chapter reports the association between the prevalence of common mental disorder, as shown by a CIS-R score of 12 or above, and a number of other factors: experience of life events, social support, and physical health difficulties. Discussion of variations in prevalence of psychiatric morbidity associated with these factors in adults across the whole age range (16–74 years) can be found in the topic report from the 2000 survey, *The Social and Economic Circumstances of Adults with Mental Disorders* (Meltzer *et al*, 2002).

3.2 Stressful life events

Other studies have shown that depressed older subjects have experienced more recent life events than non-depressed subjects (Murphy, 1982). However, older samples differ from younger ones in that chronic difficulties are more prevalent than life events, and events typically carry relatively low levels of threat (Davies, 1994a; Davies, 1994b). In older age, health difficulties are an important source of adversity (Murphy, 1982; Davies, 1994a). In a community survey in the United States the onset of illness affecting a subject or relative, the onset of money problems, and becoming a victim of crime were among the most common and most undesirable life events affecting older people (Murrell *et al*, 1984). Deaths of spouses, children and siblings had a big impact, of course, but were generally infrequent. The strength of evidence for a relationship between depression in older age and life events varies depending on the methods used to assess life events. Linn found a small significant difference in the mean number of events experienced by depressed and non-depressed community respondents (Linn *et al*, 1980). Other studies have not replicated this finding (Smallegan 1989; Murrell *et al*, 1991; Cutrona *et al*, 1986; Hurwicz *et al*, 1992). In the longitudinal Gospel Oak survey (Prince *et al*, 1997), at initial assessment depressed older people reported more life events than did non-depressed subjects. However, at follow-up (Prince *et al*, 1998) only a weak non-significant trend towards an increased risk for the onset of depression was observed. Having something important lost or stolen was the only individual event associated with the onset of depression over one year.

In this survey, respondents were shown cards listing a range of twelve stressful life events and asked to say which, if any, they had ever experienced. For every event that a respondent reported experiencing at some time in their lives they were also asked whether they had experienced that event within the preceding six months.

In addition to being asked about lifelong experience of stressful events, respondents were asked whether they had spent time in an institution, such as a children's home, borstal or young offenders unit (private education boarding schools were excluded) as a child (up to 16 years of age). They were also asked if they had spent time in local authority care.

3.2.1 Childhood experience of institutional or local authority care

Only a small proportion of respondents had spent time in an institution as a child or had been in local authority care. Although respondents who had been in an institution or in local authority care were more than twice as likely as those who had not to have a CIS-R score at or above the threshold of 12, this difference was not statistically significant. *(Table 3.1)*

3.2.2 Experience of stressful life events

Lifetime experience

Not surprisingly, given the age group covered in this report, almost everyone had experienced at least one of the stressful life events considered in this report at some time. The likelihood of having mental health problems increased with increasing

Figure 3.1 Proportion of people with a CIS-R score of 12 or above by lifetime experience of threatening life events and sex

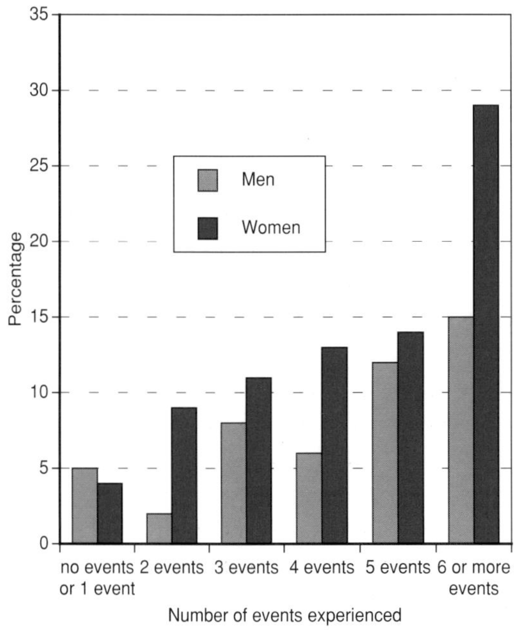

number of life events experienced among both men and women. While around one in twenty respondents reporting up to two life events had a CIS-R score of 12 or more (5% of those reporting none or one and 6% of those reporting two), this rose to one in eight (13%) of those reporting five life events, and to one in five (20%) of those reporting six or more. *(Table 3.2 and Figure 3.1)*

The first group of events that respondents were asked about related to relationship problems, illness and bereavement. They covered: serious illness, injury or assault themselves; serious illness, injury or assault of someone close to them; death of a close relative; death of a close friend or other relative; separation or divorce; and a serious problem with a close friend or relative. Given the age of the respondents, it was not surprising that most had experienced the death of someone close to them but other events had been experienced by fewer than half of all respondents.

In general, experience of these events was associated with higher levels of neurotic symptoms, but this did not always reach statistical significance. Both men and women in the sample were more likely to have a neurotic disorder if they had personally experienced serious injury or assault (14% compared with 7%) or if they had experienced separation or divorce (18% compared with 9%) than if they had not. Women were also more likely to have CIS-R scores on or above the threshold of 12 if they had experienced a serious problem with a close friend or relative (18% compared with 9%). *(Table 3.2)*

Respondents were also asked about experience of five potentially stressful events relating to employment or financial circumstances: being made redundant or sacked; looking for work for over a month; having a financial crisis; having problems with the police that led to a court appearance; losing or having stolen something of value. These less commonly occurring events were experienced by fewer than one-third of respondents in each case.

Only experience of a major financial crisis was associated with a statistically significant increase in likelihood of reporting significant levels of neurotic symptoms. Respondents who had experienced a financial crisis were three times as likely as those who had not to have a CIS-R score of 12 or more (29% compared with 9%). *(Table 3.2)*

Experience in the last 6 months

More than three-quarters of respondents had not experienced a single life-event from the list in the six months preceding the interview. However, there was a direct relationship between common mental disorder and the number of events experienced in the six months prior to interview. While fewer than one in ten respondents (9%) who had not experienced a single event had a CIS-R score of 12 or more, this increased to one in seven respondents (14%) who had experienced one event, and one in four respondents (25%) who had experienced two or more events in the preceding six months. *(Table 3.3)*

The experience of individual events in the six months preceding interview was so rare that, in most cases, it was not possible to evaluate their impact individually on the probability of having significant levels of neurotic symptoms. Events which occurred with sufficient frequency for their effect on mental health to be examined were: serious illness, injury or assault to self; serious illness, injury or assault to a close friend or relative; death of a close relative; death of a close friend or other relative; and serious problems with a close friend or relative. Although, with the exception of

'serious illness, injury or assault to a close friend or relative', those experiencing the event were more likely to have a CIS-R score of 12 or more, none of the differences tested was statistically significant.
(Table 3.3)

3.3 Social support

There is an extensive research literature which indicates that social relationships can have powerful effects on physical and mental health. However, social support has been infrequently investigated in older community samples, and the findings have not all been consistent. Bowling reported a positive association between the number of confidants and well-being in a London sample, which was not apparent in a sample drawn from a smaller town in Essex (Bowling *et al*, 1991). In both samples, health status accounted for more of the variance in well-being than did participants' social support or network characteristics. Bowling's finding of no association between social participation and well-being was contradicted by Palinkas' report in a Californian study of inverse associations between participation in church and other community organisations and depression (Palinkas *et al*, 1990a). Palinkas also reported an inverse relationship between social network size and depression. Both Palinkas and Woo, in a study from Hong Kong (Woo *et al*, 1994), report inverse associations between the frequency of contacts with friends and depression. Each of these studies adjusted to some extent for participants' physical health status, and Palinkas also controlled for age and gender. These variables are potential confounders of an association between social support and late-life depression.

There are large differences between the social support networks of older men and women, women typically having more supportive and extensive networks of friends than men (Fischer & Phillips, 1982; Palinkas *et al* 1990a). However, research tends to suggest that the that any beneficial effect of social support on the maintenance of psychological well-being tends to be stronger in men than in women (Kawachi and Berkman, 2001; Seeman, 2000).

Other studies have reported that social networks are more restricted in the presence of increasing age (Palinkas *et al*, 1990a) and disablement (Fitzpatrick *et al*, 1988; Reed *et al*, 1983). One of the more consistent findings from the literature appears to be the salience to late-life depression of contact with friends, in particular intimate, confiding relationships. A review of recent research in the area concludes that interaction with friends, rather than frequency of contact with relations, is positively associated with emotional well-being (Bowling. 1994). This view was borne out in a longitudinal study in London. No contact with friends was the only social support variable prospectively associated with the onset of depression (Prince *et al*, 1998); lack of social support and social participation were more evidently associated with the maintenance rather than with the onset of depression.

In the survey on which this report is based, several questions addressed respondents' level of social support from different perspectives. These included questions about:

- 'perceived social support', i.e. a subjective appraisal of the help, care, and affection the respondent felt they received;
- 'size of primary support group', i.e. the number of people the respondent felt close to; and
- 'number of people seen (or spoken to) in past week', i.e. a more objective index of the level of social contact respondents had.

Perceived social support

Perceived social support was assessed using a seven-item questionnaire as included in the 1992 'Health Survey for England' (Breeze *et al*, 1994) and the ONS (OPCS) surveys of psychiatric morbidity among adults in private households and in institutions catering for people with a mental disorder (Meltzer *et al*, 1995; Meltzer *et al*, 1996; Meltzer *et al*, 2002). Each question takes the form of a statement about availability of a particular aspect of social support, eg "There are people I know among my family and friends who do things to make me happy", "There are people I know among my family and friends who give me support and encouragement". Respondents are asked to rate each statement as, "Not true", "Partly true", or "Certainly true", for them. Scores of 1–3 are obtained for each question so that total scores range from 7–21. The maximum score of 21 indicates no lack of perceived social support; scores of 18–20 indicate a moderate lack, and scores of 17 and below suggest a severe lack of social support.

There was a trend for increasing rate of mental health problems with lower levels of perceived social support among both men and women, particularly among the latter group, among whom the difference between respondents with a severe lack and no lack of perceived social support was statistically significant. Among respondents severely lacking social support, 15% had a CIS-R score of 12 or more, compared with 11% of respondents moderately lacking support, and 10% of respondents not lacking social support.

(Table 3.4)

Size of primary support group

The size of respondents' primary support group, i.e. the number of friends and relatives they felt close to, was calculated from responses to questions, adapted from those used in other ONS (OPCS) surveys of psychiatric morbidity, (see Meltzer *et al*, 2002 for further detail). Previous research has suggested that adults with a primary support group of three or fewer people are at greatest risk of psychiatric morbidity (Brugha *et al*, 1987, Brugha *et al*, 1993).

Respondents with a small primary support group, (three or fewer people), were most likely, and those with a larger primary support group, (nine or more people), were least likely, to have CIS-R scores of 12 or more, (15% compared with 9%). However, these differences were not statistically significant.

(Table 3.4)

Number of friends seen in past week

To obtain an index of social isolation, respondents were asked how many of the people they felt close to or regarded as good friends (and who did not live with them) they had met or spoken with in the past week. Responses were banded into: none, one or two, and three or more. The likelihood of having a CIS-R score of 12 or more did not vary significantly between these bands. *(Table 3.4)*

3.4 Physical health

Many studies have commented on the strength of the cross-sectional relationship between physical health variables and depression in older age. There have been suggestions from clinical populations of specific associations between late-life depression, and diseases such as stroke (Eastwood *et al*, 1989) and Parkinson's disease (Cummings, 1992). However, in community surveys stroke, respiratory disease and arthritis were all found to be associated by most studies which have assessed individual diagnoses (Palinkas *et al*, 1990b; Murrell *et al*, 1983; Berkman *et al*, 1986; Lindesay, 1990; Beekman, 1996) while hypertension and diabetes were less salient. This pattern of association would suggest that disablement, the limiting long-term consequences of disease, may be more relevant than any particular pathology. Indeed, the strongest reported associations have generally been between depression and summary measures of disablement (Prince, 1998). However, these strong associations might not reflect a causal relationship. Bias (somatic contamination of the measurement of depression), confounding and, in particular reverse causality (depression leading to disablement, rather than vice versa) were plausible alternative explanations. Later prospective population-based research has clarified the association. At least five longitudinal studies have now shown a very strong association between disablement at baseline and the subsequent onset of depression (Phifer and Murrell, 1986; Kennedy *et al*, 1990; Beekman *et al*, 1995, Prince *et al*, 1998, Schoevers *et al*, 2000).). In the North London Gospel Oak survey (Prince *et al*, 1998), after adjusting for confounders, the most restricted quarter of the older population (London Handicap Scale) were six times more likely to be depressed at follow-up than the least restricted quarter.

In the survey reported on here, respondents were asked if they had any long-standing illness, disability or infirmity, and if they had they were asked what was wrong with them. Each health problem or symptom mentioned was then coded into groupings based on the International Classification of Disease (ICD-9). The conditions respondents were affected by were extremely diverse, and a summary indication of the severity of their health problems was needed. One of the levels of classification in the ICD classificatory system is in terms of the organ systems affected, and the number of organ systems affected is used as a crude indication of the extent of chronic physical health problems experienced by an individual.

Respondents of both sexes who said they had a longstanding physical health problem were much more likely than those without a physical health problem to have a CIS-R score of 12 or more, (14% compared with 3%). This association between physical health problems and neurotic symptoms was especially pronounced in men. Just one in a hundred men (1%) without a physical health problem had CIS-R scores of 12 or more compared with one in eight men (12%) with a physical health problem. Among women the equivalent figures were 5% and 16%. *(Table 3.5)*

The likelihood of having mental health problems increased with increasing extent of physical health problems. Respondents of both sexes who had no physical health problems were significantly less likely than those with one or more organ systems affected to have a CIS-R score of 12 or more. Only one in a hundred men (1%) with no physical health problems, compared with about one out of ten men with one (10%) or two (11%) organ systems affected, and one out of five (20%) of men with three or more organ systems affected, had a CIS-R score on or above the threshold. Similarly, among women, the proportion with CIS-R scores of 12 or more rose from one out of twenty women (5%) with no physical health problems, to one in four women (26%) with three or more organ systems affected. *(Table 3.5 and Figure 3.2)*

Figure 3.2 Proportion of people with a CIS-R score of 12 or above by extent of long-standing physical health problems and sex

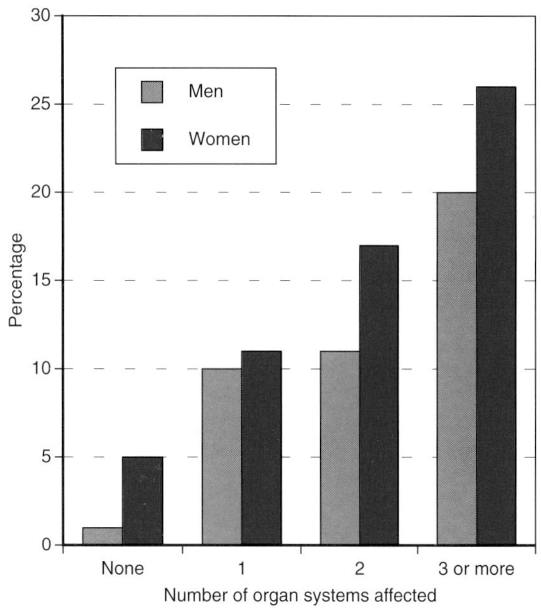

To provide an indicator of more acute illness, respondents were also asked whether they had cut down on their usual activities due to injury or illness during the two weeks preceding the interview. Respondents of both sexes who had cut down on their usual activities were at least four times as likely as those who had not to have a CIS-R score of 12 or more. Just 6% of all respondents who had not cut down on activities, compared with 29% of those who had, were found to have a CIS-R score on or above the threshold. *(Table 3.5)*

References

Beekman A T F, Deeg D J H, Smit J H and Van Tilburg W (1995) Predicting the course of depression in the older population: results from a community-based study in the Netherlands. *Journal of Affective Disorders* **34**, 41–49.

Beekman A T F (1996) Physical Health and Depression in Later Life, in *Depression in Later Life: Studies in the Community*, pp. 92–117, Hilversum:

Berkman L F, Berkman C S, Kasl S, Freeman D H *et al* (1986) Depressive Symptoms in Relation to Physical Health and Functioning in the Elderly. *American Journal of Epidemiology* **124**, 372–388.

Bowling A (1994) Social networks and social support among older people and implications for emotional well-being and psychiatric morbidity. *International Review of Psychiatry* **6**, 41–58.

Bowling A, Farquhar M and Browne P (1991) Life Satisfaction and Associations with Social Network and Support Variables in Three Samples of Elderly People. *International Journal of Geriatric Psychiatry* **6**, 549–566.

Breeze E, Maidment A, Bennett N, Flatley J and Carey S (1994) *Health Survey for England 1992*, HMSO: London.

Brown G W and Harris T O (1978) *Social Origins of Depression: A Study of Psychiatric Disorder in Women*, Tavistock Publications: London.

Brugha T, Bebbington P E, MacCarthy B, Potter J, Sturt E and Wykes T (1987) Social networks, social support and the type of depressive illness. *Acta Psychiatrica Scandinavica* **76**, 664–73.

Brugha T S, Wing J K, Brewin C R, MacCarthy B and Lesage A (1993) The relationship of social network deficits with deficits in social functioning in long-term psychiatric disorders. *Social Psychiatry and Psychiatric Epidemiology* **28**, 218–24.

Cummings J L (1992) Depression and Parkinson's disease: a review. *American Journal of Psychiatry* **149**, 443–454.

Cutrona C E, Russell D and Rose J (1986) Social support and adaption to stress by the elderly. *Journal of Psychology and Aging* **1**, 47–54.

Davies A D M (1994a) Life Events in the Normal Elderly, in Copeland J R M, Abou-Saleh, M T and Blazer D G (eds), *Principles and Practice of Geriatric Psychiatry*, John Wiley: Chichester, pp 106–113.

Davies A D M (1994b) Life Events in the Normal Elderly, in Copeland J R M, Abou-Saleh, M T and Blazer D G (eds), *Principles and Practice of Geriatric Psychiatry*, John Wiley: Chichester, pp 114–115.

Eastwood M R, Rifat S L, Nobbs H and Ruderman J (1989) Mood disorder following cerebrovascular accident. *British Journal of Psychiatry* **154**, 195–200.

Fischer C S and Phillips S L (1982) Who is alone? Social characteristics of people with small networks, in Peplau L A and Perlman D (eds) *Loneliness: a source book of current theory, research and therapy*, Wiley: New York.

Fitzpatrick R, Newman S, Lamb R and Shipley M (1988) Social relationships and psychological well-being in rheumatoid arthritis. *Social Science and Medicine* **27**, 399–403.

Hurwicz, M-L, Durham C G, Boyd-Davis S L, Gatz, M *et al* (1992) Salient life events in three generation families. *Journal of Gerontology* **47**, 11–13.

Kawachi I and Berkman L F (2001) Social ties and mental health. *Journal of Urban Health* **78**, 453–467.

Kennedy G J, Kelman H R and Thomas C (1990) The emergence of depressive symptoms in late life: the importance of declining health and increasing disability. *Journal of Community Health* **15**, 93–104.

Lindesay J (1990) The Guy's/ Age Concern Survey: Physical Health and Psychiatric Disorder in an Urban Elderly Community. *International Journal of Geriatric Psychiatry* **5**, 171–178.

Linn M W, Hunter K and Harris R (1980) Symptoms of depression and recent life events in the community elderly. *Journal of Clinical Psychology* **36**, 675–683.

Meltzer H, Gill B, Petticrew M and Hinds K (1995) *OPCS Surveys of Psychiatric Morbidity in Great Britain, Report 3: Economic activity and social functioning of adults with psychiatric disorders*, HMSO: London.

Meltzer H, Gill B, Petticrew M and Hinds K (1996) *OPCS Surveys of Psychiatric Morbidity in Great Britain, Report 6: Economic activity and social functioning of residents with psychiatric disorders*, HMSO: London

Meltzer H, Singleton N, Lee A, Bebbington P, Brugha T, Jenkins R (2002) *The Social and Economic Circumstances of Adults with Mental Disorders*, TSO: London.

Murphy E (1982) Social Origins of Depression in Old Age. *British Journal of Psychiatry* **141**, 135–142.

Murrell S A, Himmelfarb S and Wright K (1983) Prevalence of depression and its correlates in older adults. *American Journal of Epidemiology* **117**, 173–185.

Murrell S A, Norris F H and Hutchins G L (1984) Distribution and Desirability of Life Events in Older Adults: Population and Policy Implications. *Journal of Community Psychology* **12**, 301–311.

Murrell S.A, Meeks S and Walker J (1991) Protective Functions of Health and Self Esteem Against Depression in Older Adults Facing Illness or Bereavement. *Psychology and Aging* **6**, 352–360.

Palinkas L A, Wingard D L and Barrett-Connor E (1990a) The biocultural context of social networks and depression among the elderly. *Social Science and Medicine* **30**, 441–447.

Palinkas L A, Wingard D L and Barrett-Connor E (1990b) Chronic illness and depressive symptoms in the elderly: a population-based study. *Journal of Clinical Epidemiology* **43**, 1131–1141.

Phifer J F and Murrell S A (1986) Etiologic factors in the onset of depressive symptoms in older adults. *Journal of Abnormal Psychology* **95**, 282–291.

Prince M, Harwood H, Blizard R, Thomas A and Mann A (1997) Gospel Oak VI. Social support deficits, loneliness and life events as risk factors for depression in old age. *Psychological Medicine* **27**, 323–332.

Prince M J (1998) The classification and measurement of disablement, with emphasis on depression, and its applications for clinical gerontology. *Reviews in Clinical Gerontology* **8**, 227–240.

Prince M J, Harwood R, Thomas A, and Mann A H (1998) A prospective population-based cohort study of the effects of disablement and social milieu on the onset and maintenance of late-life depression. Gospel Oak VII. *Psychological Medicine* **28**, 337–350.

Reed D, McGee D, Yano K and Feinleib M (1983) Social networks and coronary heart disease among Japanese men in Hawaii. *American Journal of Epidemiology* **117**, 384–396.

Seeman T E (2000) Health promoting effects of friends and family on health outcomes in older adults. *American Journal of Health Promotion* **14**, 362–370.

Schoevers R A, Beekman A T, Deeg D J *et a.* (2000) Risk factors for depression in later life; results of a prospective community based study (AMSTEL). *Journal of Affective Disorders* **59**, 127–137.

Smallegan M (1989) Level of depressive symptoms and life stresses for culturally diverse older adults. *Gerontologist* **29**, 45–50.

Woo J, Ho S C, Lau J, Yuen Y K *et al* (1994) The prevalence of depressive symptoms and predisposing factors in an elderly Chinese population. *Acta Psychiatrica Scandinavica* **89**, 8–1.

Table 3.1 CIS-R score (grouped)

by childhood experience of institutional or local authority care

	Childhood experience of institution		Childhood experience of local authority care		Childhood experience of institution or local authority care		
	No	Yes	No	Yes	No	Yes	All
	%	%	%	%	%	%	%
CIS-R score							
Under 12	90	77	90	84	90	79	90
12 and over	10	23	10	16	10	21	10
Base	*1931*	*50*	*1928*	*40*	*1909*	*60*	*2007*

Table 3.2 Proportion of people with a CIS-R score of 12 or more by lifetime experience of threatening life events by sex

		Men	Women	All	Men	Women	All
		Percentage with CIS-R score of 12 and over			*Base*	*Base*	*Base*
Lifetime experience of threatening events							
No events or 1 event		5	4	5	*54*	*92*	*146*
2 events		2	9	6	*137*	*244*	*381*
3 events		8	11	10	*214*	*324*	*538*
4 events		6	13	9	*169*	*237*	*406*
5 events		12	14	13	*119*	*136*	*255*
6 or more events		15	29	20	*165*	*91*	*256*
Lifetime Experience of:							
Serious illness, injury or assault (to self)	No	6	9	7	*458*	*705*	*1163*
	Yes	11	18	14	*401*	*419*	*820*
Serious illness, injury or assault of close other	No	8	11	9	*604*	*759*	*1363*
	Yes	10	15	12	*255*	*365*	*620*
Death of close relative	No	6	6	6	*68*	*44*	*112*
	Yes	9	12	11	*791*	*1080*	*1871*
Death of close friend or other relative	No	8	14	11	*224*	227	451
	Yes	9	12	10	*635*	897	1532
Separation or divorce	No	7	10	9	*679*	*893*	*1572*
	Yes	16	21	18	*180*	*231*	*411*
Serious problem with close friend or relative	No	8	11	9	*750*	*938*	*1688*
	Yes	14	20	17	*109*	*186*	*295*
Made redundant or sacked	No	7	12	10	*455*	*925*	*1380*
	Yes	10	14	11	*403*	*200*	*603*
Looking for work for 1 month or longer	No	7	12	10	*637*	*1062*	*1699*
	Yes	11	16	12	*221*	*63*	*284*
Major financial crisis	No	6	11	9	*753*	*1061*	*1814*
	Yes	23	39	29	*105*	*64*	*169*
Problem with police and court appearance	No	8	12	10	*785*	*1098*	*1883*
	Yes	11	[10]	16	*73*	*27*	*100*
Something valued lost or stolen	No	8	12	10	*676*	*884*	*1560*
	Yes	10	14	12	*182*	*241*	*423*

Table 3.3 Proportion of people with a CIS-R score of 12 or more

by experience of threatening life events in the past 6 months*

		Percentage with a CIS-R score of 12 and over	*Base*
Number of threatening events experienced in the past 6 months			
No events		9	*1540*
1 event		14	*383*
2 or more events		25	*59*
Experience in the past 6 months of :			
Serious illness, injury or assault (to self)	No	10	*1933*
	Yes	22	*50*
Serious illness, injury or assault of close other	No	10	*1933*
	Yes	6	*50*
Death of close relative	No	10	*1886*
	Yes	16	*97*
Death of close friend or other relative	No	10	*1785*
	Yes	15	*198*
Serious problem with close friend or relative	No	10	*1929*
	Yes	23	*54*

* Events experienced by fewer than 30 people are not shown in the table.

Table 3.4 Proportion of people with a CIS-R score of 12 or more

by measures of social support by sex

	Men	Women	All	*Men*	*Women*	*All*
	% with CIS-R score of 12 and over			*Base*	*Base*	*Base*
Perceived social support						
Severe lack of social support	12	21	15	*95*	*56*	*151*
Moderate lack of social support	9	14	11	*203*	*213*	*416*
No lack of social support	8	11	10	*550*	*852*	*1402*
Size of primary support group						
3 or fewer people	12	20	15	*55*	*34*	*89*
4 to 8 people	9	17	13	*200*	*289*	*489*
Nine or more people	8	10	9	*599*	*798*	*1397*
Number of friends seen or spoken to in past week						
None	13	15	14	*56*	*35*	*91*
One or two	10	18	14	*160*	*173*	*333*
Three or more	8	11	10	*625*	*914*	*1539*

Table 3.5 Proportion of people with a CIS-R score of 12 or more by physical health and sex

	Men	Women	All	*Men*	*Women*	*All*
	% with CIS-R score of 12 and over			*Base*	*Base*	*Base*
Has longstanding physical illness						
No	1	5	3	*293*	*384*	*677*
Yes	12	16	14	*576*	*751*	*1327*
Number of organ systems affected by longstanding physical illness						
None	1	5	3	*293*	*384*	*677*
1	10	11	11	*323*	*385*	*708*
2	11	17	14	*163*	*237*	*400*
3 or more	20	26	23	*90*	*129*	*219*
Cut down on activities due to injuries or illness (in past 2 weeks)						
No	4	7	6	*699*	*894*	*1593*
Yes	26	32	29	*171*	*240*	*411*

4 Impact of common mental disorder

4.1 Introduction

This chapter explores the extent to which mental health problems have an impact on the functioning of older people, and on their use of health and community social services. Both functioning, and use of services are strongly influenced by age and physical health status, which in turn, as has been shown, are associated with mental health problems. Therefore it is important to identify the impact of mental health problems over and above that caused by physical health problems or other factors related to ageing. To this end, this chapter firstly considers the association between common mental disorder, as measured by the CIS-R, and mental and physical functioning, as measured by the SF-12, taking account of any effect of levels of physical health problems and age. Next the impact of common mental disorder on an individual's ability to carry out a range of Activities of Daily Living (ADLs) is considered, and finally the variation in use of a range of services between people with and without common mental disorder is covered. These analyses again take account of the effects of physical health problems and age.

Previous research has suggested that older depressed people living in one area of London were more likely to have visited their general practitioner, to have attended hospital, to have required a district nurse, to have attended a day centre, and to have the assistance of a home help (Livingston *et al*, 1990). The costs of excess service provision in the United Kingdom for older depressed persons were estimated as far back as 1986 to amount to 184 million pounds for the Health Service and 163 million pounds for social services (Smith *et al*, 1995).

A major drawback of such 'cost of illness' estimates is that they assume that any excess use of services associated with a condition is caused by that condition. This may not be the case where a third factor, such as impaired physical health, associated with both depression and service use, is implicated. The best evidence for an independent effect of depressive illness in older people comes from one comprehensive US study of health care costs suggesting a 50% increment for those with significant depressive symptoms, even after adjusting for physical illness (Unutzer *et al*, 1997).

The measure of the extent of physical health problems used in this chapter is based on the question that asked people about any longstanding physical health problems that they had. These were then coded as described in chapter 3. Since people sometimes described several symptoms relating to the same condition, people were then grouped according to the number of organ systems affected. The number of organ systems affected gives a crude but useful indication of the extent of physical health problems experienced by an individual (Burvill *et al*, 1990).

4.2 SF-12 mental and physical health summary scores

The SF-12 is a short series of questions that was derived from the SF-36, and is, "... a multipurpose short-form (SF) generic measure of health status..." (Ware *et al*, 1998). As the name suggests, it is comprised of 12 items which include a general assessment of health status and items that measure the extent to which physical and mental health problems restrict an individual's regular activities. The SF-12 is scored to generate two summary scores – one for mental health, (the MCS-12), and one for physical health, (the PCS-12). Each of the mental health and physical health summary scores are derived from responses to all 12 items but with different regression weights applied. The scoring system results in potential scores ranging between 0 and 100. Higher scores indicate better health. In this survey the standard recall version of the SF-12 was used, asking respondents to reflect on their functioning over the preceding four weeks.

Women had slightly lower MCS-12 scores than men but there was no statistically significant difference in PCS-12 scores. The mean MCS-12 score for women was 51.2, and for men was 53.1, while the mean physical health summary score was

44.8 for women and 45.3 for men. Mean mental and physical health component scores were lower (indicating poorer health) for those who had mental health problems as indicated by a CIS-R scores of 12 or more than among those who did not, among both men and women. The mean MCS-12 score for all respondents with a CIS-R score below the threshold for common mental disorder was 53.6, while that for respondents with scores of 12 or more was 38.7. The corresponding values of the PCS-12 score were 46.4 compared with 33.4. *(Table 4.1)*

Similarly, mean SF-12 scores were lower among both men and women who had a neurotic disorder than among those who had no neurotic disorder. The mean MCS-12 score was 53.7 among respondents with no neurotic disorder compared with 39.5 among respondents with any neurotic disorder. The corresponding PCS-12 scores were 46.4 compared with 34.7. *(Table 4.1)*

As might be expected, mean SF-12 mental and physical health component scores decreased significantly with increasing levels of physical health problems. Respondents with no long-term physical health problems had a mean MCS-12 score of 54.2, and mean scores declined to 52.1 for those with one organ system affected, 50.7 for those with two, and 48.1 for those with three or more organ systems affected. The corresponding mean values for the PCS-12 score were 53.0, 44.4, 40.0, and 31.4. *(Table 4.2)*

As shown in chapter 3, there is a strong association among older people between impaired physical and mental health. Therefore it is possible that the association between mental health problems and impaired functioning as shown by SF-12 scores is due to the fact that they are both associated with physical health problems. However, at every level of physical ill-health (number of organ systems affected), mean MCS-12 scores were lower among respondents with mental health problems as shown by a CIS-R score of 12 or above than among those without. For example, among respondents with one organ system affected by long-term physical problems, mean MCS-12 scores were 38.6 for those with a CIS-R score of 12 or more compared with 53.6 for those with lower CIS-R scores. Mean PCS-12 scores were also lower in respondents with CIS-R scores on or above the threshold of 12 than among those with lower scores at every level of physical illness. However, the difference was less marked than for mental functioning (MCS-12 scores) and was only statistically significant among respondents with one or two organ systems affected by long-term physical problems. This suggests that the mental health problems have an additional impact on both mental and physical functioning over and above any effect related to physical health status and that, not surprisingly, this is greater for mental health functioning. *(Table 4.2, Figure 4.1)*

Figure 4.1 Mean SF-12 Mental and physical health summary scores by CIS-R score at different levels of physical health problems

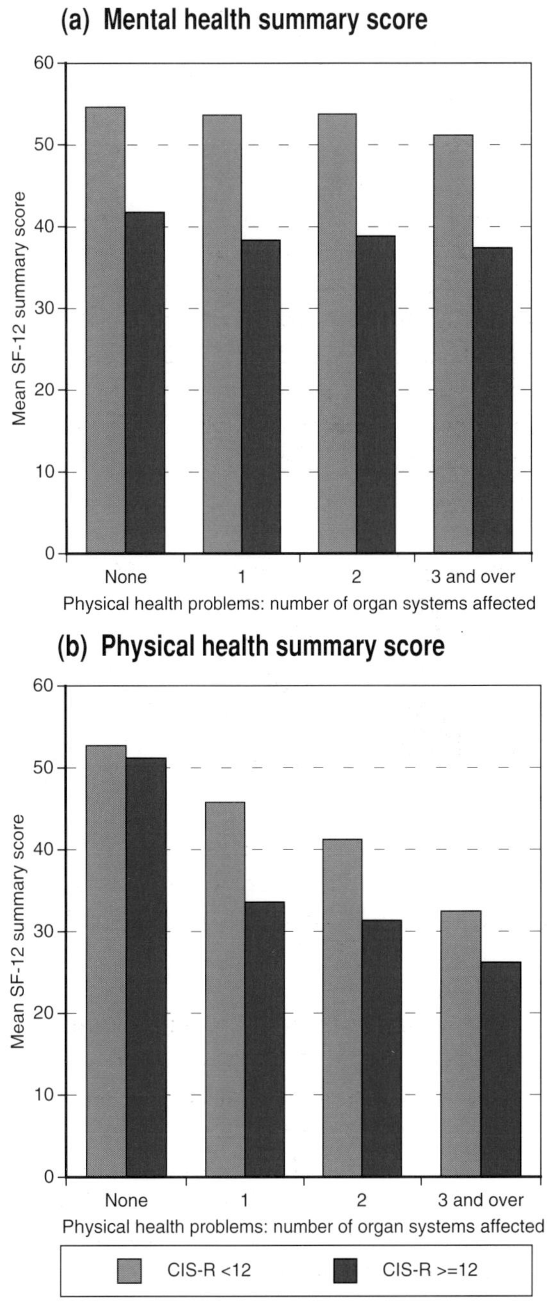

Figure 4.2 Mean SF-12 Mental and physical health summary scores by CIS-R score and age

(a) Mental health summary score

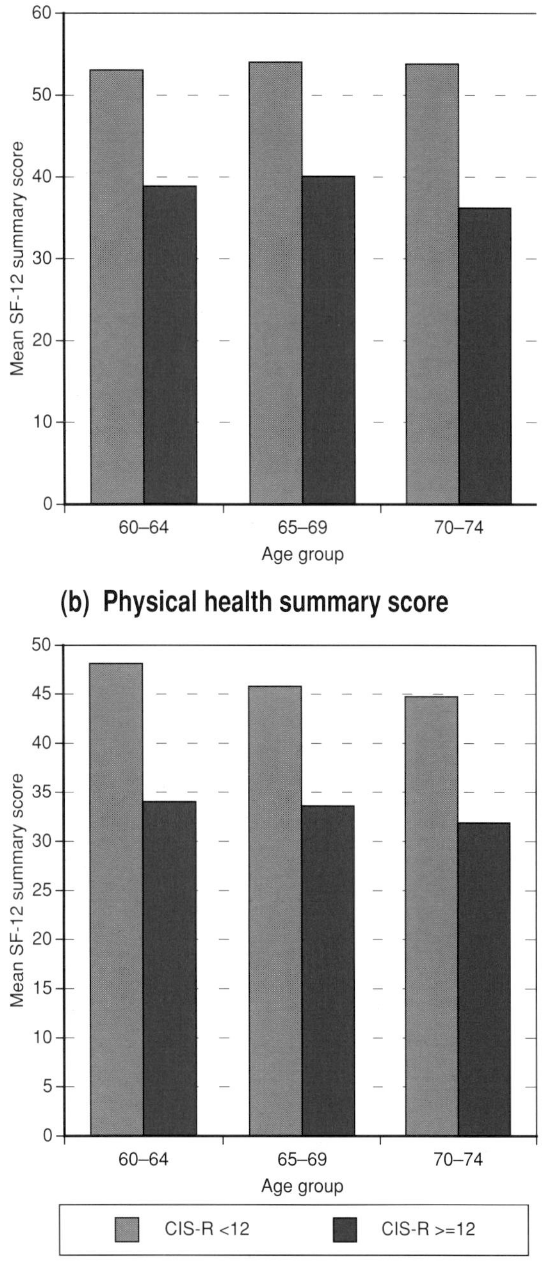

Consistent with the information on variation in distribution of neurotic symptoms and neurotic disorders with age, people aged 60 to 64 had slightly poorer mental functioning as shown by lower mean MCS-12 scores than older respondents. Also, as might be expected given the association between age and decline in physical health, mean PCS-12 scores were highest in the 60 to 64 age group and lowest in the 70 to 74 year age group. However, both mean MCS-12 and PCS-12 scores were lower among respondents with mental health problems at every age group. For example, in the 60 to 64 year age group respondents with a CIS-R score below 12 had a mean MCS-12 score of 53.1, while those with a CIS-R score of 12 or more had a mean score of 38.9. The corresponding mean PCS-12 scores were 48.1 compared with 34.0. This suggests that there is also an adverse impact of mental health problems on mental and physical functioning over an above any effect of age.

(Table 4.3, Figure 4.2)

4.3 Difficulties with activities of daily living (ADLs)

Respondents were asked a set of questions to establish whether they had any difficulties with the activities of daily living listed below.

- Personal care, such as dressing, washing, bathing, using the toilet.
- Using transport to get out and about.
- Medical care, taking medicines and pills, having injections or changing dressings.
- Household activities, such as preparing meals, shopping, laundry and housework.
- Practical activities, like gardening, decorating and doing household repairs.
- Dealing with paperwork, for example, writing letters, sending cards or filling in forms.
- Managing money – budgeting for food and paying bills.

The selection of activities was influenced by the topics covered in the MRC Needs for Care and Assessment (Brewin and Wing, 1989) and the OPCS Surveys of Disability (Martin *et al*, 1988). The seven areas of functioning chosen for this survey were the same as those included in the previous, 1993, ONS survey of psychiatric morbidity among adults (Meltzer *et al*, 1995).

4.3.1 Number of ADL difficulties

Over a third of all respondents (37%) reported having difficulty with one or more of the seven activities of daily living (ADLs), and almost half of these respondents (18% overall) reported difficulties with just one ADL. As might be expected, the proportion of respondents experiencing difficulty with any ADL was higher in this sample of older people than it was in the whole psychiatric morbidity among adults survey sample, where the figure was 21% (Meltzer *et al*, 2002).

(Table 4.4)

The likelihood of reporting difficulties with at least one ADL rose steadily with increasing CIS-R score. Respondents with few, if any, neurotic symptoms (CIS-R scores 0 to 5) were very much less likely than any other respondents to have difficulty with any ADL: 28% compared with 60% of respondents with CIS-R scores between 6 and 11, 70% of respondents with scores between 12 and 17, and 78% of respondents with scores of 18 and over. Respondents with a CIS-R score of 12 or more were also more likely than those with a lower CIS-R score to have difficulty with three or more ADLs.
(Table 4.4)

However, physical health problems will also have a big impact on ability to carry out many ADLs. As would be expected, the proportion of respondents reporting difficulty with any ADL increased with increasing number of organ systems affected by long-term physical health problems. Additionally, however, at every level of physical ill-health respondents with mental disorder, as measured by the CIS-R, were more likely than those without to have difficulty with at least one ADL. For example, among respondents with one organ system affected by physical illness, 35% of those with a CIS-R score below 12 compared with 66% of those with a CIS-R score of 12 or more had difficulties with at least one ADL. Among respondents with three or more organ systems affected by physical illness the corresponding figures were 65% compared with 94%. Furthermore, at every level of physical ill-health respondents with CIS-R scores indicative of mental disorder were more likely to have difficulty with three or more ADLs. Therefore impaired mental health appears to have an adverse impact on ability to carry out activities of daily living over and above that associated with poor physical health.
(Table 4.5, Figure 4.3)

At each age group, respondents with mental disorder were more likely than those without to have difficulty with at least one ADL. For example, among respondents aged 60 to 64, 27% of those with a CIS-R score below 12 compared with 64% of those with a CIS-R score of 12 or more had difficulties with at least one ADL. Among respondents aged 65 to 69 the corresponding figures were 34% compared with 83%, and among those aged 70 to 74, 41% compared with 81%. Therefore the adverse impact of impaired mental health upon activities of daily living seems to be independent of age. *(Table 4.6, Figure 4.4)*

Figure 4.3 Proportion of people experiencing difficulty with at least one ADL by CIS-R score at different levels of physical health problems

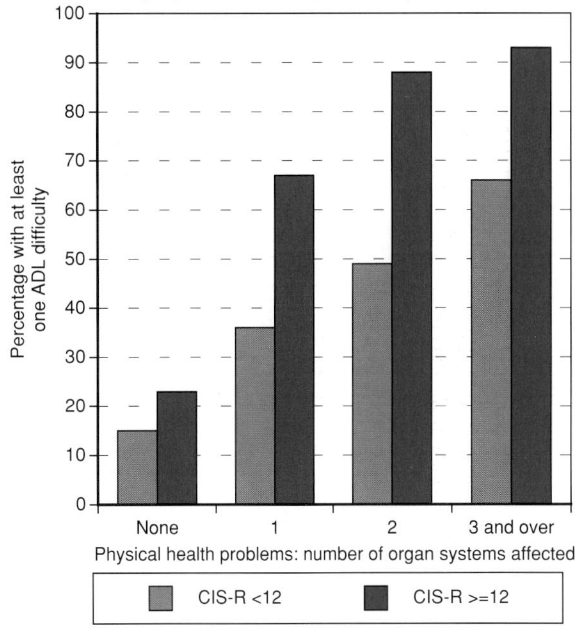

4.3.2 Type of ADL difficulty

Across all respondents the ADL with which people most frequently reported difficulty was 'practical activities' (29% of respondents had difficulty with this). The next most common sources of difficulties were 'dealing with paperwork' (14%), 'household activities' (13%), and 'using transport' (13%). Almost one in ten respondents (9%) had difficulties with 'personal care', but just 3% had difficulties 'managing money' and just 2% had difficulties with 'medical care'. This order, in terms of relative likelihood of causing difficulty, of ADLs among the older respondents covered in this report was very similar to that found among the whole survey sample. However, all ADLs except for 'managing money' were two to three times more likely to be a source of difficulty among the older sample than among the whole survey sample (Meltzer *et al*, 2002).

Respondents with mental health problems were more likely than respondents without to have difficulty with every individual ADL. Those with CIS-R scores of 12 or above were around five times as likely as those with lower CIS-R scores to report difficulties with: household activities (44% compared with 9%); managing money (10% compared with 2%); personal care (29% compared with 6%); medical care (6% compared with 1%).

Figure 4.4 Proportion of people experiencing difficulty with at least one ADL by CIS-R score and age

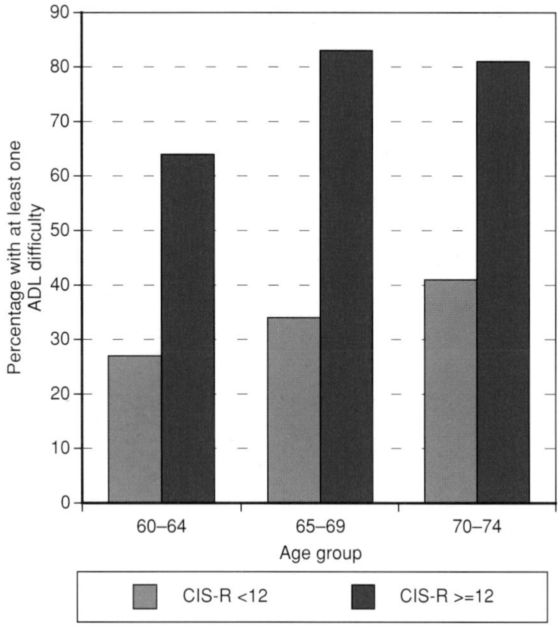

Those with CIS-R scores on or above the threshold were more than four times as likely as those with scores below 12 to have difficulties with using transport (42% compared with 10%), nearly three times as likely to have difficulties dealing with paperwork, and more than twice as likely to have difficulties with practical activities. *(Table 4.4)*

Likelihood of experiencing difficulties with each of the different ADLs tended to increase with increasing level of physical health problems (number of organ systems affected). However, at every level of physical ill-health, respondents with a CIS-R score of 12 or more were more likely than respondents with lower CIS-R scores to experience difficulty with all individual ADLs. There was less variation in the proportions reporting difficulties with individual ADLs between the different age groups but here again in each age group people with mental health problems were more likely than those without to report difficulties. *(Table 4.5 and Table 4.6)*

4.4 Use of services

While mental health problems will be related to the use of services that specifically relate to mental or emotional problems, it is possible that such problems also affect the use of services which people access primarily for physical health problems. A series of questions were asked which looked at use of services in different settings and over different time periods. People who reported use of services were asked if they had used them because of physical problems, or due to mental or emotional problems, or both. In this section, first there is a discussion of the distribution of service use and the variation in this with common mental disorder. Then, subsequent sections evaluate the contribution of common mental disorder to the use of services primarily for physical health problems independently of the effects of physical ill-health and of age. This is done by reviewing the distribution of service use among respondents with and without mental health problems, first looking at variation with different levels of physical health problems, and secondly looking at variation with age.

The four indices of service use examined were:

- contact with a GP about physical health problems and about mental health problems in the year before interview;
- outpatient or day-patient hospital visit for physical health problems and for mental health problems in the quarter prior to interview;
- inpatient hospital stay for physical health problems and for mental health problems in the 3 months prior to interview; and
- use of a range of community care services in the year before the interview, (psychiatrist, psychologist, community psychiatric nurse, community learning difficulty nurse, other nursing services, social worker, self-help/ support group, home help/home care worker, outreach worker)

4.4.1 Variation in service use with common mental disorder

Community based service use in the year prior to interview

Of all the community-based services assessed, the most commonly used was the General Practitioner service. Almost three-quarters of all respondents (72%) had consulted their GP or family doctor with a physical health problem in the year before interview, and about one in twelve (8%) had consulted for a mental health problem. Only 6% of all respondents had used any other community care service in the year prior to interview, and two-thirds of these (4%) had used a nursing service (excluding psychiatric or learning difficulty nursing). One percent of respondents in each case had seen a community psychiatric nurse, a social worker, a self-help or support group, or a home help or home care worker in the year before interview.

While overall prevalence of use of community services was low, respondents with mental health problems were more likely than those with no mental health problems to use most services assessed. This applied to services ostensibly for physical as well as mental health problems. People with CIS-R scores of 12 or more were more likely than those with lower CIS-R scores to have consulted a GP about a physical health problem in the year prior to interview, (90% compared with 70%), and more likely to have consulted a GP about a mental health problem, (30% compared with 5%). They were also three times as likely as people with lower CIS-R scores to have received some other form of community care, (15% compared with 5%). Looking at specific types of community care, as might be expected, people with mental health problems were more likely than those without such problems to have seen a community psychiatric nurse in the year prior to interview, (3% compared with less than one per cent). *(Table 4.7)*

Hospital service use in the quarter prior to interview

A quarter of all respondents (25%) had had an outpatient or day-patient hospital visit in the quarter preceding the interview, but this had been for a mental health problem in fewer than one per cent of cases. Only one in twenty-five of all respondents (4%) had had an inpatient hospital stay in the quarter preceding the interview, and again this had been for a mental health problem in fewer than one per cent of cases. Although hardly any hospital visits were specifically for mental health problems, people with mental health problems were more likely than those without such problems to have had hospital outpatient or inpatient treatment in the quarter prior to interview. In total, 44% of people with a CIS-R score over 12 had made an outpatient visit compared with 23% of those with lower CIS-R scores. The comparative figures for any inpatient visits were 9% of respondents with a CIS-R score at or above the threshold, and 4% of respondents with CIS-R scores below 12. *(Table 4.7)*

4.4.2 Impact of common mental disorder on service use at different levels of physical health

Community based service use in the year prior to interview

Likelihood of having consulted a GP about a physical health problem, but also about a mental health problem, tended to increase with increasing level of physical health problems. At every level of physical ill-health, respondents with mental health problems were more likely than those without such problems to have visited their GP about a physical health problem in the year prior to interview. For example, among respondents with one organ system affected by longstanding physical health problems, 88% of those with a CIS-R score of 12 or more, compared with 74% of those with lower CIS-R scores, had consulted their GP. The difference did not reach statistical significance at every level of physical health problems, and decreased in magnitude as the level of physical health problems increased; but rates of GP consultation were evidently very high among those with three or more complaints, regardless of mental health status. *(Table 4.8)*

At every level of physical health problems, people assessed as having a mental health problem at the time of interview were much more likely than those without such problems to have consulted their GP about a mental or emotional health problem, and the magnitude of this difference was stable across illness levels. For example, among respondents with 2 organ systems affected by physical health problems, 28% of those with a CIS-R score of 12 or more compared with just 7% of those with lower CIS-R scores had consulted their GP about a mental health problem in the year prior to interview. *(Table 4.8)*

The likelihood of having received any other community care in the year prior to interview also increased with increasing level of physical health problems and at every level of physical health problems, likelihood of having received any other community care was higher among respondents with mental health problems than among those without such problems, although the differences did not consistently reach statistical significance. For example, among respondents with two organ systems affected by physical health problems, almost one out of five (20%) of respondents with a

CIS-R score of 12 or more compared with just 6% of those with lower CIS-R scores had received other community care services in the year prior to interview. *(Table 4.8)*

Hospital service use in the quarter prior to interview

Since fewer than one percent of respondents had received inpatient or outpatient hospital treatment for a mental health problem, the analyses here are limited to hospital treatment for physical health problems. Likelihood of having hospital outpatient or inpatient treatment tended to increase with increasing level of physical health problems. Respondents with mental health problems were more likely than those without mental health problems to have had both hospital outpatient and inpatient treatment at every level of physical health problems, but these differences were not consistently statistically significant. As an example of the differences observed, among respondents with one organ system affected by physical health problems, 41% of those with a CIS-R score of 12 or more compared with just 24% of those with lower CIS-R scores had received hospital outpatient treatment for a physical health problem. *(Table 4.8)*

4.4.3 *Impact of common mental disorder on service use at different ages*

Community based service use in the year prior to interview

Having mental health problems was associated with a greater likelihood of having consulted a GP, for both mental and physical health problems, in every age group. For example, among respondents aged 60 to 64 years, 88% of those with a CIS-R score of 12 or more, compared with 66% of those with lower CIS-R score had consulted a GP for a physical health problem. Similarly, 33% compared with 6% respectively had consulted for a mental health problem. *(Table 4.9)*

The likelihood of having received any other community care in the year prior to interview was also consistently higher among people with mental health problems than among those without such problems across all age groups, although the differences were not all statistically significant. For example, among respondents aged 70 to 74 years, 23% of those with a CIS-R score of 12 or more compared with 6% of those with lower CIS-R scores had received some form of community care, apart from GP services, in the year prior to interview. *(Table 4.9)*

Hospital service use in the quarter prior to interview

As in the previous section on hospital visits, the extremely low frequency of hospital visits for mental health problems, means that the analyses described here refer only to hospital visits for physical health problems. In the 60 to 64 years and 65 to 69 years age groups, people with mental health problems were more likely than those without such problems to have had outpatient treatment for a physical health problem. For example, in the 60 to 64 year age group, almost half (49%) of the respondents with a CIS-R score above 12 had received outpatient treatment for a physical health problem in the quarter preceding interview compared with fewer than one out of five (19%) of the respondents with lower CIS-R scores. *(Table 4.9)*

The likelihood of having had a hospital inpatient stay did not vary consistently between respondents with and without mental health problems across age groups, and was generally low. Only in the 65–69 year age group, was the difference between people with and without mental health problems statistically significant. In this age group, 15% of respondents with a CIS-R score of 12 or more compared with just 4% of respondents with lower CIS-R scores had received inpatient treatment. *(Table 4.9)*

These results suggest that mental health problems are associated with increased use of community health and social care services, and hospital outpatient services, and that this pattern of increased service use occurs across different levels of physical health problems and different ages. The association between mental health problems and hospital admissions is less clear cut. However, it should be noted that the nature of the survey upon which this report is based means that it is not possible to evaluate how appropriate this level of service use is.

References

Brewin C R, Wing J K, Mangen S P *et al* (1987) Principles and practice of measuring needs in the long-term mentally ill: the MRC needs for care assessment. *Psychological Medicine*, **17**, 971–981.

Burvill P W, Mowry B and Hall W D (1990) Quantification of Physical Illness in Psychiatric Research in the Elderly. *International Journal of Geriatric Psychiatry* **5**, 161–170.

Livingston G, Thomas A, Graham N, *et al* (1990) The Gospel Oak Project: the use of health and social services by dependent elderly people in the community. *Health Trends* **2**, 70–73.

Martin J, Meltzer H and Elliot D (1988) *The OPCS Surveys of Disability in Great Britain, Report 1, The prevalence of disability among adults*, HMSO: London.

Meltzer H, Gill B, Petticrew M and Hinds K (1995) *OPCS Surveys of Psychiatric Morbidity in Great Britain. Report 2: Physical complaints, service use and treatment of adults with psychiatric disorder*, HMSO: London.

Meltzer H, Singleton N, Lee A, Bebbington P, Brugha T and Jenkins R (2002) *The Social and Economic Circumstances of Adults with Mental Disorders*, TSO: London.

Smith K, Shah A, Wright K *et al* (1995) The prevalence and costs of psychiatric disorders and learning disabilities. *British Journal of Psychiatry* **166**, 9–18.

Unutzer J, Patrick D L, Simon G *et al* (1997) Depressive Symptoms and the Cost of Health Services in HMO Patients Aged 65 Years and Older. *Journal of the American Medical Association* **277**, 1618–1623.

Ware J E, Kosinski M and Keller S (1998) SF-12:How to score the SF-12 physical and mental health summary scales, Lincoln, RI: Quality Metric Incorporated, Third Edition.

Table 4.1 Mean SF–12 mental and physical health summary scores by CIS–R score (grouped), neurotic disorder and sex

	SF–12 Mental Health Summary Score								
	Men		Women		All		*Men*	*Women*	*All*
	Mean	(s.d.)	Mean	(s.d.)	Mean	(s.d.)	*Base*	*Base*	*Base*
CIS–R score									
0–5	55.1	(6.4)	54.5	(6.9)	54.8	(6.6)	*697*	*816*	*1513*
6–11	49.3	(8.4)	45.2	(9.2)	46.9	(9.1)	*97*	*178*	*275*
Under 12	**54.4**	**(6.9)**	**52.9**	**(8.2)**	**53.6**	**(7.6)**	***794***	***994***	***1788***
12–17	41.5	(9.1)	42.0	(9.9)	41.8	(9.6)	*42*	*78*	*120*
18 and over	34.9	(8.1)	34.9	(11.1)	34.9	(10.0)	*35*	*64*	*99*
12 and over	**38.4**	**(9.3)**	**38.9**	**(11.0)**	**38.7**	**(10.4)**	***77***	***142***	***219***
Neurotic disorder									
Any anxiety disorder	..	..	39.8	(11.2)	39.6	(10.6)	*28*	*61*	*89*
Any neurotic disorder excluding mixed anxiety and depressive disorder	37.5	(10.2)	38.9	(11.4)	38.3	(11.0)	*38*	*75*	*113*
Any neurotic disorder	**39.2**	**(9.4)**	**39.6**	**(11.0)**	**39.5**	**(10.4)**	***84***	***161***	***245***
No neurotic disorder	**54.5**	**(6.9)**	**53.0**	**(8.1)**	**53.7**	**(7.5)**	***787***	***975***	***1762***
All	53.1	(8.4)	51.2	(9.7)	52.1	(9.1)	*871*	*1136*	*2007*
	SF–12 Physical Health Summary Score								
	Mean	(s.d.)	Mean	(s.d.)	Mean	(s.d.)	*Base*	*Base*	*Base*
CIS–R score									
0–5	48.1	(10.0)	47.4	(10.6)	47.7	(10.3)	*697*	*816*	*1513*
6–11	36.1	(13.7)	40.6	(14.2)	38.8	(14.2)	*97*	*178*	*275*
Under 12	**46.6**	**(11.2)**	**46.1**	**(11.7)**	**46.4**	**(11.5)**	***794***	***994***	***1788***
12–17	33.5	(12.6)	35.4	(13.5)	34.6	(13.2)	*42*	*78*	*120*
18 and over	29.9	(10.3)	33.4	(12.0)	32.0	(11.5)	*35*	*64*	*99*
12 and over	**31.8**	**(11.7)**	**34.5**	**(12.9)**	**33.4**	**(12.5)**	***77***	***142***	***219***
Neurotic disorder									
Any anxiety disorder	..	..	40.3	(13.4)	38.7	(13.8)	*28*	*61*	*89*
Any neurotic disorder excluding mixed anxiety and depressive disorder	33.7	(13.0)	39.8	(13.3)	37.4	(13.5)	*38*	*75*	*113*
Any neurotic disorder	**32.3**	**(12.0)**	**36.1**	**(13.6)**	**34.7**	**(13.1)**	***84***	***161***	***245***
No neurotic disorder	**46.7**	**(11.2)**	**46.1**	**(11.6)**	**46.4**	**(11.4)**	***787***	***975***	***1762***
All	45.4	(12.0)	44.8	(12.4)	45.1	(12.2)	*871*	*1136*	*2007*

Table 4.2 Mean SF-12 mental and physical health summary scores by CIS-R score (grouped) and extent of physical health problems

Number of organ systems affected by longstanding physical health problems	CIS-R Score	MCS-12 Mean	MCS-12 (s.d)	PCS-12 Mean	PCS-12 (s.d.)	*Base*
None	Under 12	54.7	(6.8)	53.0	(5.7)	*651*
	12 and over	..	..	..	..	*26*
	All	54.2	(7.4)	53.0	(5.7)	*677*
One	Under 12	53.6	(7.5)	45.6	(10.8)	*630*
	12 and over	38.6	(10.5)	34.0	(12.4)	*78*
	All	52.1	(9.1)	44.4	(11.5)	*708*
Two	Under 12	52.7	(8.1)	41.4	(11.8)	*337*
	12 and over	38.3	(11.6)	31.1	(9.5)	*63*
	All	50.7	(10.0)	40.0	(12.0)	*400*
Three or more	Under 12	51.1	(9.3)	32.7	(12.3)	*167*
	12 and over	38.1	(8.6)	26.8	(9.0)	*52*
	All	48.1	(10.6)	31.4	(11.9)	*219*
All	Under 12	53.6	(7.6)	46.4	(11.5)	*1788*
	12 and over	38.7	(10.4)	33.4	(12.5)	*219*
	All	52.1	(9.1)	45.1	(12.2)	*2007*

Table 4.3 Mean SF-12 mental and physical health summary scores by CIS-R score (grouped) and age

Age	CIS-R Score	MCS-12 Mean	MCS-12 (s.d)	PCS-12 Mean	PCS-12 (s.d.)	*Base*
60–64	Under 12	53.1	(7.7)	48.1	(10.9)	*638*
	12 and over	38.9	(9.8)	34.0	(12.5)	*101*
	All	51.3	(9.3)	46.3	(12.0)	*739*
65–69	Under 12	54.0	(7.7)	45.8	(11.8)	*603*
	12 and over	40.1	(10.6)	33.6	(13.3)	*65*
	All	52.8	(8.9)	44.7	(12.4)	*668*
70–74	Under 12	53.8	(7.4)	44.7	(11.5)	*547*
	12 and over	36.2	(11.1)	31.9	(11.4)	*53*
	All	52.4	(9.1)	43.7	(12.0)	*600*
All	Under 12	53.6	(7.6)	46.4	(11.5)	*1788*
	12 and over	38.7	(10.4)	33.4	(12.5)	*219*
	All	52.1	(9.1)	45.1	(12.2)	*2007*

Table 4.4 Difficulties with activities of daily living

by CIS–R score (grouped)

	CIS–R score						
	0–5	6–11	12–17	18 and over	Under 12	12 and over	All
	%	%	%	%	%	%	%
Number of ADL difficulties							
0	72	40	30	22	67	27	63
1	16	27	13	23	18	17	18
2	6	10	13	7	6	10	7
3	4	10	16	17	5	16	6
4 or more	3	13	29	31	5	30	7
Any ADL difficulty	**28**	**60**	**70**	**78**	**33**	**73**	**37**
Base	*1493*	*273*	*118*	*96*	*1766*	*214*	*1980*
	Percentage reporting each difficulty						
Type of ADL difficulty							
Practical activities	20	50	57	70	25	63	29
Dealing with paperwork	10	20	34	34	12	34	14
Household activities	7	21	44	43	9	44	13
Using transport	8	20	41	44	10	42	13
Managing money	2	3	9	11	2	10	3
Personal care	4	17	30	28	6	29	9
Medical care	1	3	6	6	1	6	2
Base	*1494*	*273*	*119*	*96*	*1767*	*215*	*1982*

Table 4.5 Difficulties with activities of daily living

by CIS–R score (grouped) and extent of physical health problems

	Number of organ systems affected by longstanding physical health problems									
	0		1		2		3 and over		All	
	CIS–R score									
	Under 12	12 and over	Under 12	12 and over	Under 12	12 and over	Under 12	12 and over	Under 12	12 and over
	%	%	%	%	%	%	%	%	%	%
Number of ADL difficulties										
0	85	[19]	65	34	52	13	35	6	67	27
1	11	[2]	20	18	25	16	21	21	18	17
2	2	[2]	6	7	10	21	13	5	6	10
3	1	[-]	4	11	8	21	16	26	5	16
4 or more	0	[3]	5	30	6	28	16	41	5	30
Any ADL difficulty	**15**	**[7]**	**35**	**66**	**48**	**87**	**65**	**94**	**33**	**73**
	Percentage reporting each difficulty									
Type of ADL difficulty										
Practical activities	7	[6]	26	51	41	73	54	90	25	63
Dealing with paperwork	9	[6]	14	39	11	39	17	29	12	34
Household activities	1	[1]	9	39	14	55	31	57	9	44
Using transport	1	[3]	11	39	13	50	32	55	10	42
Managing money	1	[3]	2	12	2	2	5	15	2	10
Personal care	0	[1]	5	24	11	29	25	48	6	29
Medical care	-	-	2	4	1	3	4	15	1	6
Base	*642*	*26*	*623*	*78*	*334*	*61*	*165*	*50*	*1767*	*215*

Table 4.6 Difficulties with activities of daily living

by CIS–R score (grouped) and age

	Age							
	60–64		65–69		70–74		All	
	CIS–R score							
	Under 12	12 and over	Under 12	12 and over	Under 12	12 and over	Under 12	12 and over
	%	%	%	%	%	%	%	%
Number of ADL difficulties								
0	73	36	66	17	59	19	67	27
1	15	14	18	23	21	15	18	17
2	5	6	6	13	8	15	6	10
3	3	16	5	16	6	17	5	16
4 or more	3	28	5	29	6	35	5	30
Any ADL difficulty	**27**	**64**	**34**	**83**	**41**	**81**	**33**	**73**
	Percentage reporting each difficulty							
Type of ADL difficulty								
Practical activities	19	57	26	67	32	70	25	63
Dealing with paperwork	10	32	11	34	14	37	12	34
Household activities	7	37	10	47	12	53	9	44
Using transport	7	39	10	41	13	51	10	42
Managing money	2	13	3	3	2	11	2	10
Personal care	5	25	6	31	8	35	6	29
Medical care	0	4	1	4	2	13	1	6
Base	*630*	*98*	*593*	*65*	*544*	*52*	*1767*	*215*

Table 4.7 Use of services

by CIS–R score (grouped)

	CIS–R score						
	0–5	6–11	12–17	18 and over	Under 12	12 and over	All
	Percentage using each service						
Service use							
Visited GP with physical complaint in last year	67	85	88	91	70	90	72
Visited GP with mental complaint in last year	4	13	20	41	5	30	8
Any GP visit in last year	**69**	**87**	**90**	**94**	**71**	**92**	**74**
Had outpatient visit for physical complaint in last quarter	21	35	41	46	23	43	25
Had outpatient visit for mental complaint in last quarter	0	0	-	2	0	1	0
Any outpatient visit in last quarter	**21**	**35**	**41**	**47**	**23**	**44**	**25**
Had inpatient visit for physical complaint in last quarter	3	6	8	9	4	8	4
Had inpatient visit for mental complaint in last quarter	-	-	-	1	-	0	0
Any inpatient visit in last quarter	**3**	**6**	**8**	**10**	**4**	**9**	**4**
Seen psychiatrist in last year	0	0	2	3	0	2	0
Seen psychologist in last year	-	-	-	2	-	1	0
Seen community psychiatric nurse in last year	0	1	1	4	0	3	1
Seen community learning difficulty nurse in last year	-	-	-	-	-	-	-
Used other nursing services in last year	3	5	5	4	4	5	4
Seen social worker in last year	0	2	3	6	1	4	1
Used self help/support group in last year	0	1	3	1	0	2	1
Used home help/home care worker in last year	1	2	2	6	1	4	1
Seen outreach worker in last year	0	0	-	-	0	-	0
Received any community care in last year	**4**	**10**	**14**	**17**	**5**	**15**	**6**
Base	*1513*	*275*	*120*	*99*	*1788*	*219*	*2007*

Table 4.8 Use of services

by CIS-R scores (grouped) and extent of longstanding physical health problems

	Number of organ systems affected by longstanding physical health problems									
	0		1		2		3 and over		All	
	CIS-R score									
	Under 12	12 and over	Under 12	12 and over	Under 12	12 and over	Under 12	12 and over	Under 12	12 and over
	Percentage using each service									
Service use										
Visited GP with physical complaint in last year	55	[16]	73	87	82	94	92	98	70	90
Visited GP with mental complaint in last year	5	[14]	4	24	7	28	7	32	5	30
Any GP visit in last year	**57**	**[21]**	**74**	**88**	**84**	**96**	**93**	**98**	**71**	**92**
Had outpatient visit for physical complaint in last quarter	13	[3]	24	41	28	42	46	61	23	43
Had outpatient visit for mental complaint in last quarter	0	[2]	0	-	0	-	0	-	0	1
Any outpatient visit in last quarter	**14**	**[5]**	**24**	**41**	**28**	**42**	**47**	**61**	**23**	**44**
Had inpatient stay for physical complaint in last quarter	2	-	5	5	4	12	8	12	4	8
Had inpatient stay for mental complaint in last quarter	-	-	-	1	-	-	-	-	-	0
Any inpatient visit in last quarter	**2**	**-**	**5**	**6**	**4**	**12**	**8**	**12**	**4**	**9**
Received any community care in last year	**3**	**[3]**	**6**	**11**	**6**	**20**	**13**	**20**	**5**	**15**
Base	*651*	*26*	*630*	*78*	*337*	*63*	*167*	*52*	*1788*	*219*

Table 4.9 Use of services

by CIS–R scores (grouped) and age

	Age							
	60–64		65–69		70–74		All	
	CIS-R score							
	Under 12	12 and over	Under 12	12 and over	Under 12	12 and over	Under 12	12 and over
	Percentage using each service							
Service use								
Visited GP with physical complaint in last year	66	88	72	92	72	89	70	90
Visited GP with mental complaint in last year	6	33	4	20	6	36	5	30
Any GP visit in last year	**68**	**91**	**73**	**94**	**73**	**93**	**71**	**92**
Had outpatient visit for physical complaint in last quarter	19	49	24	44	27	29	23	43
Had outpatient visit for mental complaint in last quarter	0	0	0	2	1	-	0	1
Had any outpatient visit in last quarter	**19**	**49**	**24**	**45**	**27**	**29**	**23**	**44**
Had inpatient stay for physical complaint in last quarter	3	5	4	15	4	5	4	8
Had inpatient stay for mental complaint in last quarter	-	-	-	-	-	2	-	0
Any inpatient visit in last quarter	**3**	**5**	**4**	**15**	**4**	**7**	**4**	**9**
Received any community care in last year	**4**	**10**	**6**	**19**	**6**	**23**	**5**	**15**
Base	*638*	*101*	*603*	*65*	*547*	*53*	*1788*	*219*

5 Distribution of cognitive function by socio-economic characteristics

5.1 Introduction

This chapter describes the variation in scores on three measures of cognitive function between groups defined by a range of demographic and socio-economic characteristics: sex, age, marital status, household size, tenure, area type, educational qualifications, social class, income and sources of income.

The measures of cognitive functioning, which are considered in turn in this chapter, are:

- the TICS-m (Plassman et al, 1995), a short measure of cognitive function which includes a word list learning task;
- a test of verbal fluency – the Animal Naming Test; and
- the National Adult Reading Test (NART) (Nelson and Willison, 1991).

These measures are described in more detail in Chapter 1.

Increased age is frequently associated with relative impairment on most tests of cognitive function: for example in community US studies of a rural population aged 65 and over (Ganguli *et al*, 1991), and a mixed urban and rural population aged 45–64 (Cerhan *et al*, 2000). In the UK Health and Lifestyle Survey, scores on a memory test were gradually and progressively lower from age groups 25–34 to 75 and over; however disparities on a test of reasoning were not evident until age 65 or greater (Huppert, 1987). The National Adult Reading Test has been found to be relatively stable with age, being more strongly associated with prior rather than current intellectual ability in older people (Crawford *et al*, 2001). Some studies show substantial attenuation of associations between age and cognitive function in more physically healthy or active groups (Hultsch *et al*, 1993; Emery *et al*, 1995), although others show little effect of adjustment for age-related health conditions (Launer *et al*, 1993).

Associations between gender and cognitive function vary according to the age group sampled and, possibly, the nature of the cognitive test administered. Severe cognitive impairment or dementia in older age groups has been found to be more common in women than men in some studies (Andersen *et al*, 1999), but not others (Unverzagt *et al*, 2001). Average cognitive test scores are often found to be increased in women (Cerhan *et al*, 2000), but this may be specific to tests of verbal memory (Huppert, 1987; Ganguli *et al*, 1991).

Educational attainment is generally found to be strongly associated (positively) with cognitive function measured later in life (Ganguli *et al*, 1991; Cerhan *et al*, 2000) and this association was not found to be mediated substantially by adverse health factors associated with lower education (Launer *et al*, 1993). The extent to which this association concerns stable 'premorbid' cognitive function or risk of declining function remains less clear. Social class estimated by occupation has also been found to be associated with cognitive function (Cerhan *et al*, 2000) and, in men, this was found to be independent of age, mood, and education (Gallacher *et al*, 1999).

When considering these results it must be remembered that many of these factors may themselves be inter-related, for example age and marital status or educational qualifications and social class. In order to identify those factors independently associated with cognitive decline, multiple logistic regression analysis was undertaken and the results are described at the end of section 5.2. Guidance on interpretation of the results are given in Chapter 2, in section 2.2.4).

5.2 The TICS-m

Four values from the TICS-m test are presented in the tables in this chapter. In addition to mean TICS-m total score, mean immediate word recall scores and mean delayed word recall score, the proportion of respondents with total TICS-m scores of less than 21 is given. A score of less than 21 is indicative of cognitive impairment.

5.2.1 *Personal and domestic characteristics*

About one out of five (19%) of all respondents aged 60 to 74 years had a TICS-m score below 21 (indicative of cognitive impairment), and there was no difference between men and women in likelihood of having a low TICS-m score. Across all respondents, mean TICS-m total score was 23.7 and, again, there was no sex difference in scores.

Although women had slightly higher immediate and delayed word recall scores than men, only the difference in delayed word recall was statistically significant. Mean immediate word recall scores were 4.6 in women and 4.4 in men, while delayed recall scores were 3.2 and 2.9, respectively.

Among both men and women, the likelihood of having a TICS-m score below 21 increased with age, and, correspondingly, mean total TICS-m scores, immediate and delayed word recall scores, decreased with increasing age. However, there were sex differences in the pattern of variation with age. Among men, the likelihood of having a low TICS-m score did not differ between the 60–64 and 65–69 years age groups, but it was significantly higher among those aged 70–74 than in the younger age groups. About one out of six men (16%) in both the 65–69 and 60–64 age groups scored below 21, but this rose to one out of four men aged 70 to 74 (25%). The same pattern was seen for mean scores. *(Table 5.1 and Figure 5.1)*

Figure 5.1 Proportion of respondents with signs of cognitive impairment by age and sex

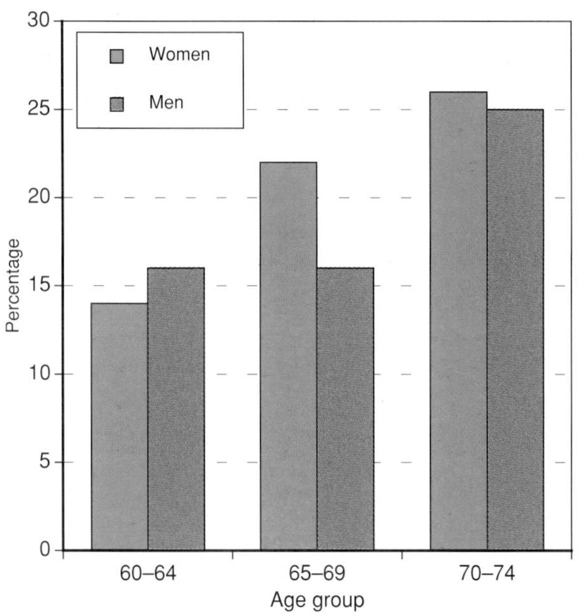

Among women, the increase in the proportion with cognitive impairment occurred at an earlier age. About one out of seven women (14%) aged 60–64 years had a total TICS-m score below 21 but this increased to over one out of five women (22%) aged 65 to 69 years, and one in four women (26%) aged 70–74 years. Correspondingly, mean total TICS-m scores decreased from 24.3 among women aged 60–64 years to 23.7 for those aged 65–69 and to 23.2 among those aged 70–74 years. Immediate and delayed word recall scores showed the same pattern of variation with age as total scores. *(Table 5.1)*

Among both men and women, widowed respondents were most likely to show signs of cognitive impairment (23% of widowed men and 26% of widowed women had a TICS-m score below 21), and those least likely to do so were divorced or separated men (16%) and single women (14%). However, only the difference in likelihood between single and widowed women reached statistical significance. Although mean TICS-m total scores, immediate and delayed recall word recall scores tended to be lowest among widowed respondents of both sexes, most differences in mean scores between respondents with different marital status were not statistically significant. It should be noted that the tendency for lower scores among widowed people may be associated with the fact that a higher proportion of them were aged 70 or over. Differences in household size were not associated with significant variation in likelihood of cognitive impairment or on TICS-m scores. *(Table 5.2)*

Respondents of both sexes who lived in rented accommodation were almost twice as likely as those living in accommodation which they owned outright or with a mortgage to show signs of cognitive impairment. Almost one in three people (30%) living in rented accommodation compared with only about one in six (16%) living in accommodation they owned had a TICS-m score below 21. Similarly, both men and women who lived in rented accommodation had significantly lower mean TICS-m total scores, immediate and delayed word recall scores than other respondents. The mean TICS-m total score was 22.3 among respondents living in rented accommodation compared with 24.2 among those who owned their home with a mortgage and 24.1 among those who owned their homes outright. *(Table 5.3)*

Men and women who lived in urban areas were more likely than those living in semi-rural, or rural areas to have low TICS-m scores. However, only the difference between men living in urban areas (21% with a low TICS-m score) and those living in semi-rural areas (14% with a low TICS-m score) was statistically significant. Mean TICS-m total scores and immediate and delayed word recall scores did not vary significantly between different types of area in either sex. *(Table 5.3)*

5.2.2 Education and social class

All TICS-m scores were strongly associated with the level of educational qualifications attained in both sexes. Likelihood of having a score below 21 decreased with increasing level of education, and mean TICS-m total scores and immediate and delayed word recall scores increased with increasing level of educational qualifications. Mean TICS-m total scores were 22.8 among people with no educational qualifications, 24.5 among those with GCSE level qualifications, and 25.7 among those with qualifications at A' level or above. *(Table 5.4 and Figure 5.2)*

Figure 5.2 Mean TICS-m scores by educational qualifications and sex

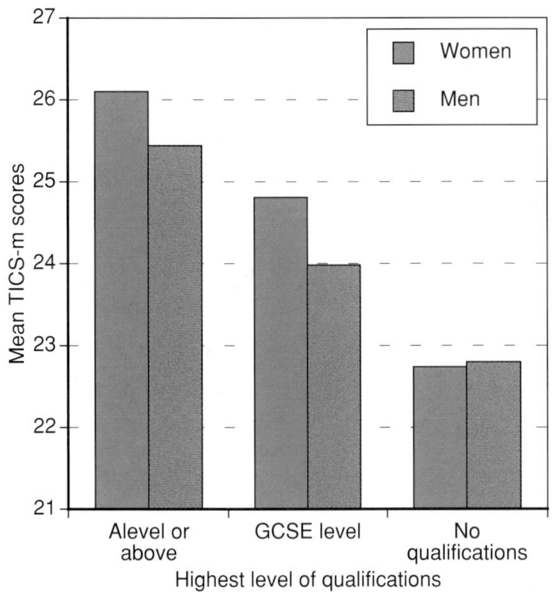

The likelihood of showing signs of cognitive impairment tended to increase with decreasing social class, and TICS-m mean total scores, immediate and delayed word recall scores decreased with decreasing social class in both sexes. Overall, there appeared to be a marked difference between those in manual and non-manual occupations. For example, while 10% of people in Social Classes I and II and 14% of those in Class III non-manual had a TICS-m score below 21 this rose to 23% of those in Class III manual, 26% of those in IV and 36% of those in Social Class V.

Social class was assigned on the basis of each individual's last occupation, so people who never worked are not included here, and those who only worked part-time to provide a supplementary income may be included in a lower group than usual. This will occur more often among women. However, the pattern of association between levels of cognitive impairment was, in general, similar among men and women, although the small group of men in Social Class IIINM had a rate similar to those in manual occupational groups. *(Table 5.4)*

5.2.3 Income

The likelihood of cognitive impairment decreased with increasing household income, and TICS-m mean scores, and immediate and delayed word recall scores increased with increasing household income, in both sexes. More than 1 in 4 people with a gross weekly household income of less than £200 had a TICS-m score below 21, and this decreased to less than 1 in 14 of those with a weekly income of £400 or more. A similar pattern was seen for both men and women. Correspondingly, mean TICS-m total scores were lower among people with a weekly household income below £200 and the highest mean score was that for people with a household income of £400 or more. For example, mean TICS-m total scores were 22.5 where income was under £100, 24.1 where income was between £200 and £299, and 25.5 in the group with and income of £500 or more. The same pattern applied to the associations between weekly household income and mean immediate and delayed word recall scores. *(Table 5.5)*

Respondents in receipt of some form of state benefits were more likely than those not receiving any benefits to have a TICS-m score below 21, (20% compared with 13%), but the difference was statistically significant only among men. However, consideration of differences between men and women is complicated by the different eligibility for benefits resulting from the difference in retirement age. A similar pattern was apparent for mean TICS-m total scores, and immediate and delayed word recall scores. Men receiving some form of state benefits had a mean TICS-m total score of 23.4 compared with a mean total score of 24.6 among men not receiving any benefits.

(Table 5.6)

The association between receipt of benefit and distribution of TICS-m scores varied between different benefit types, aspects of the TICS-m, and between the sexes. Receipt of housing benefit was consistently associated with greater likelihood of cognitive impairment. Both men and women living in households that received housing benefit were more likely than the sample as whole to have a TICS-m score below 21, (34% compared with 19%). This group also had lower mean total TICS-m, immediate and delayed word recall scores than the sample overall. In general, people receiving an income-related benefit were more likely to have indications of cognitive impairment. For example, men receiving income-related benefits were more likely than men in general to have a TICS-m score below 21, (32% compared with 18%) and women receiving any these benefits had lower mean TICS-m total scores than women in general, (22.0 compared with 23.8). Across men and women together, those receiving a disability-related benefit also had lower mean TICS-m total and immediate word recall scores than people in general, but the difference was less marked.

(Table 5.6)

Among the population covered by this report, receipt of state benefits is not as directly linked to income levels as is the case in younger age groups because of the high proportion receiving the State pension. Respondents were therefore grouped according to whether or not they had sources of income instead of, or as well as, benefits. Both men and women whose only source of income was state benefits were more likely than other people to have a TICS-m score below 21, (30% compared with 17% of respondents with income from benefit plus other sources, and 12% of respondents with non-benefit income only). A similar pattern was seen for mean TICS-m total scores and immediate and delayed word recall scores with lowest mean scores among respondents, of both sexes, whose only source of income was state benefits and highest among those who had exclusively non-benefit income.

(Table 5.7)

5.2.4 Variables independently associated with TICS-m scores

To examine their independent association with cognitive impairment, a logistic regression analysis of TICS-m scores on: sex, age, marital status, household size, nature of tenure, type of area (urban, semi-rural, rural), level of highest educational qualifications, social class, household income, receipt of disability related state benefits, receipt of income related state benefits, receipt of housing benefit, receipt of any other type of state benefit, was carried out. (These were the same variables as using the logistic regression analysis described in Chapter 2). Table 5.8 presents the results of the analysis, indicating the socio-economic characteristics that were independently associated with TICS-m scores and the odds ratios associated with each category of these socio-economic characteristics. In this case the 'odds' under consideration is the ratio of the proportion of respondents with a TICS-m score below 21, (i.e. indicative of cognitive impairment), to the proportion of respondents having higher TICS-m scores.

Four of the characteristics examined were independently associated with cognitive impairment: age, level of highest educational qualifications, social class, and household income. As expected the odds of having cognitive impairment increased with age. People aged 70–74 years had higher odds than did people in the reference group, aged 60–64 years (OR 1.66). The odds of having cognitive impairment also increased with decreasing educational attainment. The odds of cognitive impairment for people with no qualifications (OR 2.61) and for those with GCSE level qualifications (OR 1.74) were higher than for people with qualifications at 'A' level or above.

(Table 5.8)

Even after taking into account the effect of educational attainment, social class was associated with TICS-m scores. People in the manual social classes (IIIM: OR 1.48; IV: OR 1.53; and V: OR 2.20) had significantly higher odds of having cognitive impairment than people in social classes I and II combined did.

Similarly, even after taking into account age, educational attainment and social class, household income was associated with TICS-m scores. People in the lowest income bands, with an income of less than £200, and those whose income was unknown, had more than double the odds of having cognitive impairment that people in the highest income band (£500 or more) had. *(Table 5.8)*

Education, social class and income therefore appeared to have additive influences on level of cognitive function in older people. Income here is likely to be measuring a dimension of socio-economic position throughout earlier life which is incompletely represented by occupational status. This may be determined by higher levels of cognitive function already achieved or may itself determine further opportunities to improve functioning. Missing data in epidemiological surveys is generally increased in people with higher levels of morbidity and this may well explain the association between unknown income and cognitive impairment.

5.3 The National Adult Reading Test (NART)

As discussed in chapter 1, the NART is a measure of 'crystallised' intelligence and as such should not be affected in the early stages of cognitive impairment, in contrast to the TICS-m. Tables 5.8 to 5.14 present both mean NART scores and mean estimated verbal IQ. As there is a direct relationship between these two sets of values, since the verbal IQ score is obtained by applying an algorithm to the raw NART score, it can be inferred that variations with demographic characteristics will be the same for both. Therefore, in the discussion below only mean NART scores are referred to.

5.3.1 *Personal and domestic characteristics*

The mean NART score for the sample as a whole was 27.8, and did not differ between men and women. There was also no variation in mean NART scores with age in either sex. *(Table 5.9)*

Single women had higher NART scores than women with any other marital status, but otherwise there was no apparent variation in NART scores with marital status. The mean NART score for single women was 35.6 compared with a mean of 27.7 among married women. People living in one- and two-person households had higher NART scores than those living in households of three or more people. Mean NART scores were 27.8 among respondents in one-person households, 28.5 among those in two-person households, and 24.1 among those in households of three or more people. *(Table 5.10)*

Among both sexes, mean NART scores were higher among people living in accommodation which they owned (with a mortgage or outright) than among those living in rented accommodation. Both men and women living in rural areas had higher mean NART scores than did their urban counterparts. Mean NART score among all rural respondents was 30.5 compared with a mean of 26.7 among urban respondents. Among men, the mean NART score of respondents living in semi-rural areas was also higher than that of men living in urban areas, (30.1 compared with 26.4). *(Table 5.11)*

5.3.2 *Education and social class*

As might be expected, there was a clear trend in both sexes for mean NART score to increase with level of educational qualifications attained. The mean NART score was 22.7 among respondents with no qualifications, this rose to 31.0 among respondents with GCSE level qualifications, and to 37.8 among respondents with qualifications at A' level or above. *(Table 5.12)*

Mean NART scores decreased with decreasing social class in both men and women. People in the non-manual Social Classes (I or II, and III non-manual) had higher mean NART scores than people in the manual Social Classes (III manual, IV, and V). People in Social Class V had significantly lower mean NART scores than all other groups. The mean NART score for social classes I or II combined was 35.0 and mean scores declined to 19.2 in Social Class V. *(Table 5.12)*

5.3.3 *Income*

Mean NART score increased with increasing weekly household income in both men and women. For example, mean NART scores were 36.7 for women with a weekly household income of £500 or more, 30.5 for those with an income between £300 and £399, and 24.8 for those with a weekly household income under £100. *(Table 5.13 and Figure 5.3)*

Figure 5.3 Mean NART scores by household income and sex

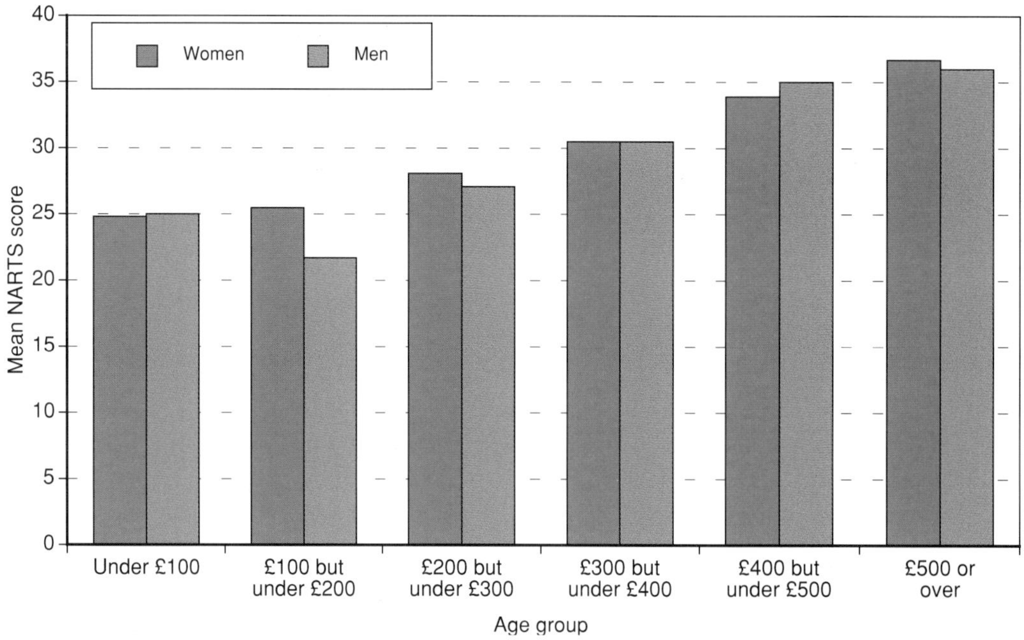

The difference in mean NART scores between respondents receiving and not receiving any state benefits was larger in women than in men, but was not statistically significant in either sex. However, both men and women who were in receipt of a disability-related benefit, an income-related benefit, or housing benefit, had significantly lower mean NART scores than respondents in general. The 'other benefit' group had generally higher mean NART scores and this group will include many people in receipt of retirement pension only. Both men and women whose only source of income was state benefits had lower mean NART scores than other people. The mean NART score was 22.2 for those dependent solely on benefits compared with 29.3 among those whose income came from a combination of state benefits and other sources and 29.0 among people whose income came entirely from other sources.
(Tables 5.14 and 5.15)

5.4 Verbal fluency

Verbal fluency (VF) was measured through the animal naming test in which people score one for every different animal they can name within one minute.

5.4.1 Personal and domestic characteristics

The mean VF score (the number of animals named in one minute), across all respondents, was 21.7. There was no sex difference in scores, (among women the mean score was 21.3 and among men it was 22.1). Verbal fluency decreased with increasing age in both men and women. Among respondents overall, the mean score ranged from 23.0 among people aged 60–64 down to 20.2 for people aged 70–74 years. *(Table 5.16)*

The pattern of variation in verbal fluency associated with marital status was different for men and women. Among women, the mean VF score was highest among those who were single (22.6) and lowest among those who were widowed (20.5), but the differences between women with different marital status were not statistically significant. Among men, the mean VF score was highest among those who were married (22.4) and lowest among those who were single (19.8), and this difference was statistically significant. There was no significant variation in mean VF scores with household size in either men or women.
(Table 5.17)

Tenure was associated with variation in verbal fluency in both men and women. Respondents living in rented accommodation had lower mean scores than those living in accommodation which they owned with a mortgage or outright, (19.3 compared with 23.8, and 22.0 respectively). Among men there was also a significant difference in mean VF scores between respondents owning their homes outright (22.1) and those owning their homes with a mortgage (24.8). This may reflect the association between increasing age and both declining VF and increased likelihood of having paid off a mortgage. (Table 5.18)

People living in rural areas had higher VF scores than did people living in urban or semi-rural areas. Across all respondents, the mean VF score was 24.0 for rural dwellers, 21.3 for those in semi-rural areas, and 21.4 for urban dwellers. (Table 5.18)

5.4.2 Education and social class

Verbal fluency increased with increasing level of educational qualifications attained in both women and men. Among women, those with no qualifications had the lowest mean VF score (19.4), and this rose to 22.5 among those with GCSE level qualifications and then to 26.3 for women with 'A' level qualifications. Men with 'A' level qualifications also had higher mean VF score than did men with no qualifications, 24.6 compared with 20.5. (Table 5.19)

Mean VF score tended to increase with social class among both men and women. People in social classes I or II combined had a higher mean VF score than people in Social Classes IV or V, (23.7 compared with 19.9 and 19.6 respectively). (Table 5.19)

5.4.3 Income

Verbal fluency was highest among men and women in the two highest weekly household income groups, (between £400 and £499, and £500 and over). Mean VF scores were lowest in respondents with a household income between £100 and £199 but there was little difference in scores between groups with incomes below £400. Respondents with a mean income of £500 or more had a mean VF score of 24.7, and those with a mean income between £100 and £199 had a mean VF score of 19.8. (Table 5.20)

Respondents in receipt of some form of state benefit had lower mean VF scores than those not receiving any benefits, but this difference was statistically significant only among men. The mean VF score for men receiving benefits was 21.1 compared with 25.0 in men not receiving any benefits. As for the other measures of cognitive status, it was those groups receiving benefits other than pensions who tended to have lower verbal fluency. For example, women receiving some form of disability-related benefit had lower mean VF scores than women in general did, (18.4 compared with 21.3), and men receiving some form of income-related benefit and men receiving housing benefit had lower mean VF scores than men in general did. Men receiving income-related benefit had a mean VF score of 18.2, for those in receipt of housing benefit it was 18.6, while the mean score for men in general which was 22.1. (Table 5.21)

There was an association between scores on the verbal fluency test and people's source of income. Those dependent solely on state benefits for their income had a lower mean VF score than did people whose income came wholly or partly from other sources. Across all respondents, the mean VF score for those dependent on state benefits was 19.1, this rose to 21.8 among those whose income came from benefits plus other sources, and to 25.3 for those whose income came entirely from other sources. This association may be in part due to age differences in the groups, as the group with non-benefit income only will be composed largely of men aged 60–65 who have not yet retired. (Table 5.22)

5.5 The association between TICS-m scores, NART scores and educational qualifications

The three measures of cognitive functioning included in this survey were included because they measure slightly different aspects of cognitive function. While the TICS-m is a screen for signs of cognitive impairment, eg problems with concentration and short term memory, the NART is a measure of crystallised intelligence, which should remain stable with age and in the early stages of cognitive decline. However, both were shown above to be strongly associated with education. To investigate the association between the two measures taking education into account, the proportion of people with signs of cognitive impairment and the mean TICS-m and immediate and delayed word recall scores for different groups characterised by educational level and NART score were considered.

The likelihood of having signs of cognitive impairment increased with decreasing NART score. Taking people with all levels of educational qualifications together, those with the lowest NART scores (0–19) were about twice as likely as those with a NART score between 20 to 29 to have a TICS-m score below 21, (33% compared with 17%), and more than five times as likely as those with the highest NART scores (40–50) to have a low TICS-m score, (33% compared with 6%). A similar pattern was seen at all levels of educational attainment. For example, among those with no educational qualifications the proportion of people with signs of cognitive impairment ranged from 35% among those with the lowest NART scores (0–19) down to 11% among the group with the highest scores (40–50).

TICS-m mean total and mean immediate and delayed word recall scores showed a similar pattern, with higher mean scores being associated with higher NART scores at every level of educational qualifications. *(Table 5.23)*

References

Andersen K, Launer L J, Dewey M E, Letenneur L, Ott A, Copeland J R M, Dartigues J-F, Kragh-Sorensen P, Baldereschi M, Brayne C, Lobo A, Martinez-Lage J M, Stijnen T and Hofman A (1999) Gender differences in the incidence of AD and vascular dementia. *Neurology* **53**, 1992–1997.

Cerhan J R, Folsom A R, Mortimer J A, Shahar E, Knopman D S, McGovern P G, Hays M A, Crum L D and Heiss G (2000) Correlates of cognitive function in middle-aged adults. Atherosclerosis Risk in Communities (ARIC) Study Investigators. *Gerontology* **44**, 95–105.

Crawford J R, Deary I J, Starr J and Whalley L J (2001) The NART as an index of prior intellectual functioning: a retrospective validity study covering a 66-year interval. *Psychological Medicine* **31**, 451–458.

Emery C F, Huppert F A and Schein R L (1995) Relationships among age, exercise, health, and cognitive function in a British sample. *Gerontologist* **35**, 378–385.

Gallacher J E, Elwood P C, Hopkinson C, Rabbitt P M, Stollery B T, Sweetnam P M, Brayne C and Huppert F A (1999) Cognitive function in the Caerphilly study: associations with age social class, education and mood. *European Journal of Epidemiology* **15**, 161–169.

Ganguli M, Ratcliff G, Huff F J, Belle S, Kancel M J, Fischer L, Seaberg E C and Kuller L H (1991) Effects of age, gender, and education on cognitive tests in a rural elderly community sample: Norms from the Monongahela Valley Independent Elders Survey. *Neuroepidemiology* **10**, 42–52.

Hultsch D F, Hammer M and Small B J (1993) Age differences in cognitive performance in later life: relationships to self-reported health and activity life style. *Journal of Gerontology* **48**, P1–P11.

Huppert F A (1987) Cognitive function, in Cox B D, Blaxter M, Buckle A L J, Fenner, N P, Golding J.F, Gore M, Huppert F A, Nickson J, Roth M, Stark J, Wadsworth M E J and Wichelow M J (eds), *Health and Lifestyle Survey: preliminary report of a nationwide survey of the physical and mental health, attitudes and lifestyle of a random sample of 9003 British adults*, London: Health Promotion Trust, pp 43–50.

Launer L J, Dinkgreve M A H M, Jonker C, Hooijer C and Lindeboom J (1993) Are age and education independent correlates of the Mini-Mental State Exam performance of community-dwelling elderly? *Journal of Gerontology* **48**, P271–P277.

Nelson H E and Willison J R (1991) *National Adult Reading Test (NART)* (Second Edition), NFER-Nelson: Windsor.
Plassman B L, Welsh K A, Helms M, Brandt J, Page W F and Breitner J C (1995) Intelligence and education as predictors of cognitive state in late life: a 50-year follow-up. *Neurology* **45**(**8**), 1446–50.

Plassman B L, Welsh K A, Helms M, Brandt J, Page W F and Breitner J C (1995) Intelligence and education as predictors of cognitive state in late life: a 50-year follow-up. *Neurology* **45**(**8**), 1446–50.

Unverzagt F W, Gao S, Baiyewu O, Ogunniyi A O, Gureje O, Perkins A, Emsley C L, Dickens J, Evans R, Musick B, Hall K S, Hui S L and Hendrie H C (2001) Prevalence of cognitive impairment. Data from the Indianapolis Study of Health and Aging. *Neurology* **57**, 1655–1662.

Table 5.1 Mean TICS-m scores and percentage of people scoring below the cut-off for cognitive impairment by age and sex

	Age							
	60–64		65–69		70–74		All	
	Percentage of respondents							
TICS-m score								
Women								
Under 21 (Cognitive impairment)	14		22		26		20	
	Mean	*(s.d)*	*Mean*	*(s.d)*	*Mean*	*(s.d)*	*Mean*	*(s.d)*
TICS-m total score	24.5	(4.2)	23.6	(4.5)	22.9	(4.4)	23.7	(4.4)
Immediate word recall score	4.8	(1.5)	4.6	(1.7)	4.3	(1.5)	4.6	(1.6)
Delayed word recall score	3.5	(1.6)	3.2	(1.6)	2.9	(1.7)	3.2	(1.7)
Base	*404*		*370*		*355*		*1129*	
	Percentage of respondents							
Men								
Under 21 (Cognitive impairment)	16		16		25		18	
	Mean	*(s.d)*	*Mean*	*(s.d)*	*Mean*	*(s.d)*	*Mean*	*(s.d)*
TICS-m total score	24.1	(3.6)	23.8	(3.8)	22.9	(3.9)	23.7	(3.8)
Immediate word recall score	4.5	(1.3)	4.4	(1.5)	4.1	(1.4)	4.4	(1.4)
Delayed word recall score	3.2	(1.4)	3.0	(1.5)	2.6	(1.5)	2.9	(1.5)
Base	*329*		*290*		*241*		*860*	
	Percentage of respondents							
All adults								
Under 21 (Cognitive impairment)	15		19		25		19	
	Mean	*(s.d)*	*Mean*	*(s.d)*	*Mean*	*(s.d)*	*Mean*	*(s.d)*
TICS-m total score	24.3	(3.9)	23.7	(4.2)	22.9	(4.2)	23.7	(4.1)
Immediate word recall score	4.7	(1.4)	4.5	(1.6)	4.2	(1.5)	4.5	(1.5)
Delayed word recall score	3.3	(1.6)	3.1	(1.6)	2.8	(1.6)	3.1	(1.6)
Base	*733*		*660*		*596*		*1989*	

Table 5.2 Mean TICS-m scores and percentage of people scoring below the cut-off for cognitive impairment by marital status, household size and sex

	Marital status				Household Size			
	Married	Divorced or Separated	Single	Widowed	1	2	3 and over	All
	Percentage of respondents							
TICS-m score								
Women								
Under 21 (Cognitive impairment)	19	18	14	26	21	21	17	20
	Mean (s.d)	*Mean (s.d)*	*Mean (s.d)*	*Mean (s.d)*	*Mean (s.d)*	*Mean (s.d)*	*Mean (s.d)*	*Mean (s.d)*
TICS-m total score	23.9 (4.4)	24.1 (4.3)	24.4 (3.9)	23.0 (4.5)	23.5 (4.3)	23.9 (4.5)	23.4 (4.3)	23.7 (4.4)
Immediate word recall score	4.7 (1.6)	4.6 (1.7)	4.5 (1.5)	4.3 (1.6)	4.4 (1.6)	4.6 (1.6)	4.6 (1.5)	4.6 (1.6)
Delayed word recall score	3.3 (1.6)	3.2 (1.6)	3.3 (1.8)	3.0 (1.7)	3.1 (1.6)	3.3 (1.7)	3.2 (1.6)	3.2 (1.7)
Base	*587*	*132*	*67*	*343*	*476*	*578*	*75*	*1129*
	Percentage of respondents							
Men								
Under 21 (Cognitive impairment)	18	16	20	23	22	16	23	18
	Mean (s.d)	*Mean (s.d)*	*Mean (s.d)*	*Mean (s.d)*	*Mean (s.d)*	*Mean (s.d)*	*Mean (s.d)*	*Mean (s.d)*
TICS-m total score	23.7 (3.8)	24.1 (3.5)	23.5 (4.1)	23.4 (4.0)	23.4 (4.0)	23.9 (3.6)	23.1 (4.5)	23.7 (3.8)
Immediate word recall score	4.4 (1.4)	4.4 (1.5)	4.2 (1.6)	4.3 (1.5)	4.2 (1.6)	4.4 (1.4)	4.4 (1.4)	4.4 (1.4)
Delayed word recall score	3.0 (1.4)	3.0 (1.6)	2.9 (1.6)	2.9 (1.6)	2.8 (1.7)	3.0 (1.4)	2.8 (1.4)	2.9 (1.5)
Base	*571*	*104*	*71*	*114*	*234*	*552*	*74*	*860*
	Percentage of respondents							
All adults								
Under 21 (Cognitive impairment)	19	17	17	25	22	18	20	19
	Mean (s.d)	*Mean (s.d)*	*Mean (s.d)*	*Mean (s.d)*	*Mean (s.d)*	*Mean (s.d)*	*Mean (s.d)*	*Mean (s.d.)*
TICS-m total score	23.8 (4.1)	24.1 (3.9)	24.0 (4.0)	23.1 (4.4)	23.5 (4.2)	23.9 (4.0)	23.2 (4.4)	23.7 (4.1)
Immediate word recall score	4.5 (1.5)	4.5 (1.6)	4.4 (1.6)	4.3 (1.5)	4.3 (1.6)	4.5 (1.5)	4.5 (1.5)	4.5 (1.5)
Delayed word recall score	3.1 (1.6)	3.1 (1.6)	3.1 (1.7)	3.0 (1.6)	3.0 (1.6)	3.2 (1.6)	3.0 (1.5)	3.1 (1.6)
Base	*1158*	*236*	*138*	*457*	*710*	*1130*	*149*	*1989*

Table 5.3 Mean TICS-m scores and percentage of people scoring below the cut-off for cognitive impairment by tenure and area type and sex

	Tenure						Area type							
	Owned Outright		Owned with mortgage		Rented		Urban		Semi-rural		Rural		All	
	Percentage of respondents													
TICS-m score														
Women														
Under 21 (Cognitive impairment)	17		17		29		22		18		17		20	
	Mean	*(s.d)*	*Mean*	*(s.d)*	*Mean*	*(s.d)*	*Mean*	*(s.d)*	*Mean*	*(s.d)*	*Mean*	*(s.d)*	*Mean*	*(s.d)*
TICS-m total score	24.3	(4.0)	24.3	(3.9)	22.2	(4.9)	23.6	(4.5)	23.9	(4.2)	24.4	(3.8)	23.7	(4.4)
Immediate word recall score	4.7	(1.6)	4.7	(1.5)	4.1	(1.6)	4.5	(1.6)	4.6	(1.5)	4.8	(1.7)	4.6	(1.6)
Delayed word recall score	3.4	(1.6)	3.4	(1.5)	2.8	(1.6)	3.2	(1.6)	3.3	(1.7)	3.1	(1.7)	3.2	(1.7)
Base	*717*		*101*		*304*		*720*		*306*		*103*		*1129*	
	Percentage of respondents													
Men														
Under 21 (Cognitive impairment)	16		16		30		21		14		17		18	
	Mean	*(s.d.)*	*Mean*	*(s.d)*	*Mean*	*(s.d)*	*Mean*	*(s.d)*	*Mean*	*(s.d)*	*Mean*	*(s.d)*	*Mean*	*(s.d)*
TICS-m total score	23.9	(3.7)	24.3	(3.5)	22.5	(4.2)	23.5	(3.9)	23.9	(3.9)	24.1	(3.2)	23.7	(3.8)
Immediate word recall score	4.4	(1.4)	4.5	(1.3)	4.0	(1.5)	4.4	(1.4)	4.3	(1.5)	4.5	(1.4)	4.4	(1.4)
Delayed word recall score	3.0	(1.4)	3.2	(1.5)	2.6	(1.5)	2.9	(1.4)	3.0	(1.6)	3.0	(1.3)	2.9	(1.5)
Base	*545*		*129*		*181*		*492*		*244*		*124*		*860*	
	Percentage of respondents													
All adults														
Under 21 (Cognitive impairment)	16		16		30		21		16		17		19	
	Mean	*(s.d)*	*Mean*	*(s.d)*	*Mean*	*(s.d)*	*Mean*	*(s.d)*	*Mean*	*(s.d)*	*Mean*	*(s.d)*	*Mean*	*(s.d)*
TICS-m total score	24.1	(3.9)	24.2	(3.7)	22.3	(4.6)	23.5	(4.2)	23.9	(4.1)	24.2	(3.5)	23.7	(4.1)
Immediate word recall score	4.6	(1.5)	4.6	(1.4)	4.1	(1.6)	4.5	(1.5)	4.4	(1.5)	4.6	(1.5)	4.5	(1.5)
Delayed word recall score	3.2	(1.6)	3.3	(1.5)	2.7	(1.6)	3.1	(1.6)	3.1	(1.6)	3.0	(1.5)	3.1	(1.6)
Base	*1262*		*230*		*485*		*1212*		*550*		*227*		*1989*	

Table 5.4 Mean TICS-m scores and percentage of people scoring below the cut-off for cognitive impairment by highest educational qualifications, social class and sex

	Highest educational qualifications			Social class					
	A level or above	GCSE level	No qualifications	I and II	III Non-manual	III Manual	IV	V	All
	Percentage of respondents								
TICS-m score									
Women									
Under 21 (Cognitive impairment)	8	12	27	13	12	27	29	34	20
	Mean (s.d.)	*Mean (s.d.)*	*Mean (s.d.)*	*Mean (s.d.)*	*Mean (s.d.)*	*Mean (s.d.)*	*Mean (s.d.)*	*Mean (s.d.)*	*Mean (s.d.)*
TICS-m total score	26.1 (3.5)	24.8 (3.8)	22.7 (4.4)	25.2 (4.0)	24.6 (3.8)	23.2 (4.0)	22.2 (4. 0)	21.7 (5.1)	23.7 (4.4)
Immediate word recall score	5.3 (1.5)	4.8 (1.5)	4.3 (1.6)	5.0 (1.6)	4.7 (1.5)	4.5 (1.4)	4.2 (1.5)	4.0 (1.6)	4.6 (1.6)
Delayed word recall score	3.9 (1.6)	3.5 (1.6)	2.9 (1.6)	3.6 (1.7)	3.4 (1.5)	3.1 (1.7)	2.8 (1.6)	2.8 (1.7)	3.2 (1.7)
Base	*173*	*289*	*663*	*263*	*402*	*108*	*202*	*126*	*1129*
	Percentage of respondents								
Men									
Under 21 (Cognitive impairment)	6	15	26	9	23	22	22	40	18
	Mean (s.d.)	*Mean (s.d.)*	*Mean (s.d.)*	*Mean (s.d.)*	*Mean (s.d.)*	*Mean (s.d.)*	*Mean (s.d.)*	*Mean (s.d.)*	*Mean (s.d.)*
TICS-m total score	25.4 (3.1)	24.0 (3.5)	22.8 (3.7)	25.0 (3.4)	23.7 (3.6)	23.0 (3.8)	22.9 (3.5)	22.1 (3.8)	23.7 (3.8)
Immediate word recall score	4.9 (1.4)	4.4 (1.4)	4.1 (1.4)	4.7 (1.5)	4.3 (1.4)	4.2 (1.3)	4.2 (1.4)	4.0 (1.2)	4.4 (1.4)
Delayed word recall score	3.4 (1.4)	3.0 (1.5)	2.7 (1.5)	3.3 (1.4)	2.6 (1.5)	2.9 (1.4)	2.6 (1.5)	2.6 (1.4)	2.9 (1.5)
Base	*213*	*187*	*455*	*309*	*68*	*303*	*121*	*46*	*860*
	Percentage of respondents								
All adults									
Under 21 (Cognitive impairment)	7	14	26	10	14	23	26	36	19
	Mean (s.d.)	*Mean (s.d.)*	*Mean (s.d.)*	*Mean (s.d.)*	*Mean (s.d.)*	*Mean (s.d.)*	*Mean (s.d.)*	*Mean (s.d.)*	*Mean (s.d.)*
TICS-m total score	25.7 (3.3)	24.5 (3.7)	22.8 (4.1)	25.1 (3.6)	24.5 (3.8)	23.1 (3.8)	22.5 (3.8)	21.7 (4.8)	23.7 (4.1)
Immediate word recall score	5.0 (1.5)	4.7 (1.4)	4.2 (1.5)	4.8 (1.5)	4.7 (1.5)	4.3 (1.4)	4.2 (1.4)	4.0 (1.5)	4.5 (1.5)
Delayed word recall score	3.6 (1.5)	3.3 (1.6)	2.8 (1.6)	3.4 (1.6)	3.3 (1.6)	2.9 (1.5)	2.7 (1.6)	2.7 (1.7)	3.1 (1.6)
Base	*386*	*476*	*1118*	*572*	*470*	*411*	*323*	*172*	*1989*

Table 5.5 Mean TICS-m scores and percentage of people scoring below the cut-off for cognitive impairment by household income and sex

	Household weekly gross income						
	Under £100	£100 but under £200	£200 but under £300	£300 but under £400	£400 but under £500	£500 or over	All
	Percentage of respondents						
TICS-m score							
Women							
Under 21 (Cognitive impairment)	29	28	16	10	5	5	20
	Mean (s.d.)	*Mean (s.d.)*	*Mean (s.d.)*	*Mean (s.d.)*	*Mean (s.d.)*	*Mean (s.d.)*	*Mean (s.d.)*
TICS-m total score	22.6 (4.6)	23.1 (4.3)	24.3 (4.4)	24.7 (4.1)	26.1 (3.6)	26.2 (3.1)	23.7 (4.4)
Immediate word recall score	4.2 (1.6)	4.3 (1.5)	4.8 (1.6)	4.9 (1.6)	5.1 (1.9)	5.4 (1.2)	4.6 (1.6)
Delayed word recall score	2.9 (1.6)	3.0 (1.7)	3.4 (1.6)	3.7 (1.6)	3.9 (1.6)	4.0 (1.6)	3.2 (1.7)
Base	*71*	*72*	*84*	*93*	*43*	*62*	*1129*
	Percentage of respondents						
Men							
Under 21 (Cognitive impairment)	23	31	16	11	8	6	18
	Mean (s.d.)	*Mean (s.d.)*	*Mean (s.d.)*	*Mean (s.d.)*	*Mean (s.d.)*	*Mean (s.d.)*	*Mean (s.d.)*
TICS-m total score	22.5 (4.0)	22.4 (3.8)	23.8 (3.4)	24.4 (3.1)	25.3 (4.0)	25.1 (3.0)	23.7 (3.7)
Immediate word recall score	4.0 (1.3)	4.1 (1.3)	4.4 (1.3)	4.5 (1.3)	4.9 (1.4)	4.8 (1.3)	4.4 (1.4)
Delayed word recall score	2.6 (1.6)	2.6 (1.4)	3.0 (1.4)	3.1 (1.4)	3.5 (1.6)	3.3 (1.3)	2.9 (1.5)
Base	*79*	*223*	*185*	*107*	*64*	*94*	*860*
	Percentage of respondents						
All adults							
Under 21 (Cognitive impairment)	27	29	16	11	7	6	19
	Mean (s.d.)	*Mean (s.d.)*	*Mean (s.d.)*	*Mean (s.d.)*	*Mean (s.d.)*	*Mean (s.d.)*	*Mean (s.d.)*
TICS-m total score	22.5 (4.4)	22.8 (4.1)	24.1 (3.9)	24.5 (3.6)	25.6 (3.9)	25.5 (3.1)	23.7 (4.1)
Immediate word recall score	4.1 (1.5)	4.2 (1.4)	4.6 (1.5)	4.7 (1.5)	5.0 (1.6)	5.0 (1.3)	4.5 (1.5)
Delayed word recall score	2.8 (1.6)	2.8 (1.6)	3.2 (1.5)	3.3 (1.5)	3.7 (1.6)	3.6 (1.4)	3.1 (1.6)
Base	*258*	*624*	*379*	*200*	*107*	*165*	*1989*

Table 5.6 Mean TICS-m scores and percentage of people scoring below the cut-off for cognitive impairment by receipt of state benefits and sex

	State Benefits Received													
	Any disability related		Any income related		Any other (eg Pensions, or child related)		Household receives housing benefit		Any benefits		No benefits		All	
	Percentage of respondents													
TICS-m score														
Women														
Under 21 (Cognitive impairment)	26		28		20		32		20		15		20	
	Mean	*(s.d.)*	*Mean*	*(s.d.)*	*Mean*	*(s.d.)*	*Mean*	*(s.d.)*	*Mean*	*(s.d.)*	*Mean*	*(s.d.)*	*Mean*	*(s.d.)*
TICS-m total score	22.8	(4.6)	22.0	(4.7)	23.8	(4.3)	21.8	(4.7)	23.7	(4.3)	24.6	(4.8)	23.7	(4.4)
Immediate word recall score	4.2	(1.4)	4.2	(1.6)	4.6	(1.6)	4.0	(1.5)	4.6	(1.6)	4.9	(1.5)	4.6	(1.6)
Delayed word recall score	2.9	(1.7)	2.9	(1.6)	3.2	(1.6)	2.7	(1.7)	3.2	(1.6)	3.6	(1.6)	3.2	(1.7)
Base	*149*		*91*		*1067*		*186*		*1086*		*37*		*1129*	
	Percentage of respondents													
Men														
Under 21 (Cognitive impairment)	23		32		20		36		20		13		18	
	Mean	*(s.d.)*	*Mean*	*(s.d.)*	*Mean*	*(s.d.)*	*Mean*	*(s.d.)*	*Mean*	*(s.d.)*	*Mean*	*(s.d.)*	*Mean*	*(s.d.)*
TICS-m total score	23.0	(3.8)	22.3	(4.9)	23.5	(3.7)	21.8	(4.2)	23.4	(3.8)	24.6	(3.3)	23.7	(3.8)
Immediate word recall score	4.3	(1.3)	4.1	(1.7)	4.3	(1.4)	3.8	(1.5)	4.3	(1.4)	4.6	(1.3)	4.4	(1.4)
Delayed word recall score	2.8	(1.4)	2.6	(1.7)	2.8	(1.5)	2.4	(1.6)	2.8	(1.5)	3.4	(1.4)	2.9	(1.5)
Base	*185*		*69*		*553*		*108*		*666*		*187*		*860*	
	Percentage of respondents													
All adults														
Under 21 (Cognitive impairment)	24		30		20		34		20		13		19	
	Mean	*(s.d.)*	*Mean*	*(s.d.)*	*Mean*	*(s.d.)*	*Mean*	*(s.d.)*	*Mean*	*(s.d.)*	*Mean*	*(s.d.)*	*Mean*	*(s.d.)*
TICS-m total score	22.9	(4.1)	22.2	(4.8)	23.7	(4.1)	22.0	(4.5)	23.6	(4.1)	24.6	(3.6)	23.7	(4.1)
Immediate word recall score	4.2	(1.4)	4.1	(1.7)	4.5	(1.5)	3.9	(1.5)	4.5	(1.5)	4.7	(1.3)	4.5	(1.5)
Delayed word recall score	2.8	(1.5)	2.8	(1.6)	3.1	(1.6)	2.6	(1.6)	3.1	(1.6)	3.4	(1.4)	3.1	(1.6)
Base	*334*		*160*		*1620*		*294*		*1752*		*224*		*1989*	

Table 5.7 Mean TICS-m scores and percentage of people scoring below the cut-off for cognitive impairment by sources of income and sex

	Sources of Regular Income							
	State benefits only		Non-state benefit income only		State benefit and other income		All	
	Percentage of respondents							
TICS-m score								
Women								
Under 21 (Cognitive impairment)	31		8		16		20	
	Mean	*(s.d)*	*Mean*	*(s.d)*	*Mean*	*(s.d)*	*Mean*	*(s.d)*
TICS-m total score	22.3	(4.6)	25.7	(3.6)	24.4	(4.0)	23.7	(4.4)
Immediate word recall score	4.1	(1.5)	5.2	(1.3)	4.7	(1.6)	4.6	(1.6)
Delayed word recall score	2.8	(1.6)	3.8	(1.7)	3.4	(1.6)	3.2	(1.7)
Base	*297*		*31*		*784*		*1129*	
	Percentage of respondents							
Men								
Under 21 (Cognitive impairment)	27		13		18		18	
	Mean	*(s.d)*	*Mean*	*(s.d)*	*Mean*	*(s.d)*	*Mean*	*(s.d)*
TICS-m total score	22.6	(4.4)	24.6	(3.3)	23.6	(3.6)	23.7	(3.8)
Immediate word recall score	4.1	(1.6)	4.6	(1.3)	4.3	(1.4)	4.4	(1.4)
Delayed word recall score	2.7	(1.6)	3.3	(1.4)	2.9	(1.4)	2.9	(1.5)
Base	*121*		*184*		*543*		*860*	
	Percentage of respondents							
All adults								
Under 21 (Cognitive impairment)	30		12		17		19	
	Mean	*(s.d)*	*Mean*	*(s.d)*	*Mean*	*(s.d)*	*Mean*	*(s.d)*
TICS-m total score	22.4	(4.5)	24.8	(3.3)	24.0	(3.9)	23.7	(4.1)
Immediate word recall score	4.1	(1.6)	4.7	(1.3)	4.6	(1.5)	4.5	(1.5)
Delayed word recall score	2.7	(1.6)	3.4	(1.4)	3.1	(1.6)	3.1	(1.6)
Base	*418*		*215*		*1327*		*1989*	

Table 5.8 Odds ratios for socio-economic characteristics associated with cognitive impairment (TICS-m score under 21)

	Adjusted odds ratios	(95% confidence interval)
Socio-economic characteristics		
Age		
60 -64	1.00	———
65-69	1.22	(0.91-1.63)
70-74	1.66 **	(1.24-2.23)
Level of highest educational qualifications		
'A' level or above	1.00	———
GCSE level or equivalent	1.74 *	(1.05-2.89)
No qualifications	2.61 **	(1.62-4.22)
Social Class		
I or II	1.00	———
III non manual	0.83	(0.55-1.25)
III manual	1.48 *	(1.01-2.17)
IV	1.53 *	(1.02-2.31)
V	2.20 **	(1.38-3.48)
Not known	2.32	(1.00-5.42)
Household Income		
Not known	2.52 **	(1.26-5.03)
Less than £100	3.05 **	(1.48-6.26)
£100 to £199	3.47 **	(1.79-6.73)
£200 to £299	1.97	(0.99-3.90)
£300 to £399	1.41	(0.66-3.01)
£400 to £499	1.25	(0.49-3.17)
£500 and over	1.00	———

Other variables not significantly associated:
Sex, marital status, type of tenure, area type,
receipt of disability related benefits, receipt of income related benefits, receipt of housing benefit, receipt of any other type of state benefits

* = p <0.05
** = p <0.01

Table 5.9 Mean NART score and mean estimated verbal IQ by age and sex

	Age						All	
	60–64		65–69		70–74			
NART score								
Women	*Mean*	*(s.d.)*	*Mean*	*(s.d.)*	*Mean*	*(s.d.)*	*Mean*	*(s.d.)*
Mean NART score	27.9	(12.0)	27.4	(12.1)	27.6	(12.4)	27.6	(12.2)
Mean estimated verbal IQ	102.2	(13.6)	101.6	(13.8)	101.8	(14.2)	101.9	(13.8)
Base	*383*		*357*		*337*		*1077*	
Men								
Mean NART score	26.8	(13.4)	28.3	(12.3)	29.3	(13. 0)	28.0	(13.0)
Mean estimated verbal IQ	100.9	(15.2)	102.7	(14.1)	103.8	(14.8)	102.3	(14.8)
Base	*306*		*280*		*229*		*815*	
All adults								
Mean NART score	27.3	(12.7)	27.8	(12.2)	28.4	(12.7)	27.8	(12.6)
Mean estimated verbal IQ	101.6	(14.4)	102.1	(13.9)	102.7	(14.5)	102.1	(14.3)
Base	*689*		*637*		*566*		*1892*	

Table 5.10 Mean NART score and mean estimated verbal IQ by marital status, household size and sex

	Marital status								Household Size							
	Married		Divorced or Separated		Single		Widowed		1		2		3 or more		All	
NART score																
Women	*Mean*	*(s.d)*	*Mean*	*(s.d)*	*Mean*	*(s.d)*	*Mean*	*(s.d)*	*Mean*	*(s.d)*	*Mean*	*(s.d)*	*Mean*	*(s.d)*	*Mean*	*(s.d)*
Mean NART score	27.7	(12.1)	28.5	(12.1)	35.6	(10.0)	25.4	(12.1)	28.0	(12.2)	28.3	(12.0)	24.0	(12.4)	27.6	(12.2)
Mean estimated verbal IQ	101.9	(13.8)	102.8	(13.7)	111.0	(11.4)	99.4	(13.7)	102.3	(13.8)	102.6	(13.6)	97.8	(14.1)	101.9	(13.8)
Base	*560*		*126*		*66*		*325*		*454*		*550*		*73*		*1077*	
Men																
Mean NART score	28.0	(13.1)	27.7	(12.0)	27.5	(12.7)	28.1	(12.8)	27.4	(12.7)	28.8	(12.6)	24.2	(14.4)	28.0	(13.0)
Mean estimated verbal IQ	102.3	(14.9)	102.0	(13.7)	101.8	(14.4)	102.4	(14.5)	101.7	(14.4)	103.2	(14.3)	98.0	(16.4)	102.3	(14.8)
Base	*541*		*102*		*64*		*108*		*220*		*531*		*64*		*815*	
All adults																
Mean NART score	27.8	(12.6)	28.1	(12.1)	31.7	(12.1)	26.1	(12.3)	27.8	(12.3)	28.5	(12.3)	24.1	(13.4)	27.8	(12.6)
Mean estimated verbal IQ	102.1	(14.4)	102.5	(13.7)	106.5	(13.8)	100.2	(14.0)	102.1	(14.0)	102.9	(14.0)	97.9	(15.2)	102.1	(14.3)
Base	*1101*		*228*		*130*		*433*		*674*		*1081*		*137*		*1892*	

Table 5.11 Mean NART score and mean estimated verbal IQ by tenure, area type and sex

	Tenure			Area type			
	Owned Outright	Owned with mortgage	Rented	Urban	Semi-rural	Rural	All
NART SCORE							
Women	*Mean (s.d)*	*Mean (s.d)*	*Mean (s.d)*	*Mean (s.d)*	*Mean (s.d)*	*Mean (s.d)*	*Mean (s.d)*
Mean NART score	29.7 (11.9)	26.9 (12.3)	22.2 (10.8)	27.0 (12.0)	27.8 (12.2)	31.5 (12.4)	27.6 (12.2)
Mean estimated verbal IQ	104.3 (13.6)	101.1 (14.0)	95.7 (12.3)	101.2 (13.7)	102.1 (13.9)	106.3 (14.1)	101.9 (13.8)
Base	*693*	*97*	*282*	*681*	*296*	*100*	*1077*
Men							
Mean NART score	28.8 (12.9)	29.3 (13.2)	23.4 (11.9)	26.4 (13.2)	30.1 (12.5)	29.9 (12.3)	28.0 (13.0)
Mean estimated verbal IQ	103.3 (14.6)	103.8 (15.0)	97.1 (13.6)	100.5 (15.0)	104.8 (14.2)	104.5 (13.9)	102.3 (14.7)
Base	*521*	*124*	*165*	*460*	*235*	*120*	*815*
All adults							
Mean NART score	29.3 (12.4)	28.2 (12.9)	22.7 (11.3)	26.7 (12.6)	28.9 (12.4)	30.5 (12.4)	27.8 (12.6)
Mean estimated verbal IQ	103.8 (14.1)	102.6 (14.6)	96.3 (12.8)	100.9 (14.3)	103.4 (14.1)	105.2 (14.0)	102.1 (14.3)
Base	*1214*	*221*	*447*	*1141*	*531*	*220*	*1892*

Table 5.12 Mean NART score and mean estimated verbal IQ

by highest educational qualifications, social class and sex

	Highest educational qualifications						Social class											
	A level or above		GCSE level		No qualifications		I and II		III Non-manual		III Manual		IV		V		All	
NART SCORE																		
Women	*Mean*	*(s.d)*	*Mean*	*(s.d)*	*Mean*	*(s.d)*	*Mean*	*(s.d)*	*Mean*	*(s.d)*	*Mean*	*(s.d)*	*Mean*	*(s.d)*	*Mean*	*(s.d)*	*Mean*	*(s.d)*
Mean NART score	38.3	(8.9)	32.6	(9.8)	22.6	(11.1)	34.8	(10.7)	30.3	(10.3)	22.1	(10.8)	22.1	(11.5)	18.8	(10.3)	27.6	(12.2)
Mean estimated verbal IQ	114.1	(10.2)	107.6	(11.1)	96.2	(12.6)	110.1	(12.3)	105.0	(11.8)	95.6	(12.3)	95.6	(13.1)	91.9	(11.7)	101.9	(13.8)
Base	*164*		*282*		*629*		*249*		*395*		*104*		*190*		*117*		*1077*	
Men																		
Mean NART score	37.5	(10.1)	28.9	(11.2)	22.8	(12.1)	35.2	(10.4)	31.1	(11.5)	22.5	(12.3)	23.3	(11.6)	20.4	(11.2)	28.0	(13.0)
Mean estimated verbal IQ	113.1	(11.6)	103.3	(12.8)	96.4	(13.7)	110.5	(11.9)	105.9	(13.1)	96.1	(14.0)	97.0	(13.2)	93.6	(12.8)	102.3	(14.8)
Base	*204*		*182*		*425*		*298*		*66*		*284*		*115*		*41*		*815*	
All adults																		
Mean NART score	37.8	(9.7)	31.0	(10.6)	22.7	(11.5)	35.0	(10.6)	30.5	(10.6)	22.4	(12.0)	22.6	(11.6)	19.2	(10.6)	27.8	(12.6)
Mean estimated verbal IQ	113.5	(11.0)	105.7	(12.1)	96.3	(13.1)	110.3	(12.1)	105.1	(12.0)	96.0	(13.6)	96.1	(13.1)	92.3	(12.1)	102.1	(14.3)
Base	*368*		*464*		*1054*		*547*		*461*		*388*		*305*		*158*		*1892*	

Table 5.13 Mean NART score and mean estimated verbal IQ by household income and sex

	Household weekly gross income													
	Under £100		£100 but under £200		£200 but under £300		£300 but under £400		£400 but under £500		£500 or over		All	
NART SCORE														
Women	*Mean*	*(s.d)*	*Mean*	*(s.d)*	*Mean*	*(s.d)*	*Mean*	*(s.d)*	*Mean*	*(s.d)*	*Mean*	*(s.d)*	*Mean*	*(s.d)*
Mean NART score	24.8	(11.9)	25.5	(12.2)	28.1	(10.9)	30.5	(10.8)	33.9	(10.7)	36.7	(9.1)	27.6	(12.2)
Mean estimated verbal IQ	98.6	(13.5)	99.5	(13.9)	102.5	(12.4)	105.2	(12.3)	109.0	(12.2)	112.2	(10.4)	101.9	(13.8)
Base	*169*		*377*		*190*		*90*		*41*		*60*		*1077*	
Men														
Mean NART score	25.0	(13.7)	21.7	(11.8)	27.1	(11.4)	30.5	(12.2)	35.0	(11.)	35.0	(10.4)	28.0	(13.0)
Mean estimated verbal IQ	98.9	(15.6)	95.2	(13.4)	101.3	(13.0)	105.1	(13.9)	110.3	(12.5)	111.4	(11.8)	102.3	(14.8)
Base	*76*		*210*		*179*		*102*		*62*		*98*		*815*	
All adults														
Mean NART score	24.9	(12.5)	24.0	(12.2)	27.6	(11.2)	30.5	(11.6)	34.6	(10.9)	36.2	(10.0)	27.8	(12.6)
Mean estimated verbal IQ	98.7	(14.3)	97.8	(13.9)	101.8	(12.7)	105.2	(13.2)	109.8	(12.4)	111.7	(11.3)	102.1	(14.3)
Base	*245*		*587*		*369*		*192*		*103*		*158*		*1892*	

Table 5.14 Mean NART score and mean estimated verbal IQ by receipt of state benefits and sex

	State Benefits Received													
	Any disability related		Any income related		Any other (eg Pensions, or child related)		Household receives housing benefit		Any benefits		No benefits		All	
NART SCORE														
Women	*Mean*	*(s.d)*	*Mean*	*(s.d)*	*Mean*	*(s.d)*	*Mean*	*(s.d)*	*Mean*	*(s.d)*	*Mean*	*(s.d)*	*Mean*	*(s.d)*
Mean NART score	22.1	(11.7)	21.9	(10.3)	27.6	(12.2)	21.3	(10.7)	27.4	(12.2)	33.4	(9.2)	27.6	(12.2)
Mean estimated verbal IQ	95.6	(13.3)	95.3	(11.7)	101.8	(13.9)	94.7	(12.2)	101.7	(13.9)	108.4	(10.5)	101.9	(13.8)
Base	*139*		*83*		*1022*		*171*		*1041*		*32*		*1077*	
Men														
Mean NART score	24.7	(13.3)	21.4	(12.2)	28.6	(12.4)	21.9	(12.4)	27.9	(12.8)	28.1	(13.4)	28.0	(13.0)
Mean estimated verbal IQ	98.6	(15.2)	94.9	(13.8)	103.0	(14.1)	95.4	(14.1)	102.2	(14.5)	102.4	(15.2)	102.3	(14.8)
Base	*169*		*63*		*532*		*94*		*632*		*177*		*815*	
All adults														
Mean NART score	23.6	(12.7)	21.6	(11.3)	28.0	(12.3)	21.6	(11.4)	27.6	(12.4)	28.9	(13.0)	27.8	(12.6)
Mean estimated verbal IQ	97.3	(14.5)	95.1	(12.8)	102.3	(14.0)	95.0	(13.0)	101.9	(14.2)	103.3	(14.7)	102.1	(14.3)
Base	*308*		*146*		*1554*		*265*		*1673*		*209*		*1892*	

Table 5.15 Mean NART score and mean estimated verbal IQ by sources of income and sex

	Sources of Regular Income							
	State benefits only		Non-state benefit income only		State benefit and other income		All	
NART SCORE								
Women	*Mean*	*(s.d)*	*Mean*	*(s.d)*	*Mean*	*(s.d)*	*Mean*	*(s.d)*
Mean NART score	22.1	(10.9)	..	..	29.5	(12.0)	27.6	(12.2)
Mean estimated verbal IQ	95.6	(12.4)	..	..	104.0	(13.7)	101.9	(13.8)
Base	*274*		*29*		*762*		*1077*	
Men								
Mean NART score	22.2	(13.9)	28.2	(13.4)	29.0	(12.2)	28.0	(13.0)
Mean estimated verbal IQ	95.8	(15.8)	102.6	(15.2)	103.5	(13.9)	102.3	(14.8)
Base	*111*		*175*		*519*		*815*	
All adults								
Mean NART score	22.2	(11.9)	29.0	(13.0)	29.3	(12.1)	27.8	(12.6)
Mean estimated verbal IQ	95.7	(13.5)	103.5	(14.8)	103.8	(13.8)	102.1	(14.3)
Base	*385*		*204*		*1281*		*1892*	

Table 5.16 Mean Verbal Fluency (animal naming) Test score

by age and sex

	Age						All	
	60–64		65–69		70–74			
Verbal Fluency Test score								
Women	*Mean*	*(s.d)*	*Mean*	*(s.d)*	*Mean*	*(s.d)*	*Mean*	*(s.d)*
Mean score	22.4	(11.7)	21.1	(11.3)	19.9	(9.3)	21.3	(11.0)
Base	*388*		*351*		*340*		*1078*	
Men								
Mean score	23.5	(14.6)	21.6	(8.3)	20.4	(6.9)	22.1	(11.1)
Base	*317*		*283*		*230*		*827*	
All adults								
Mean score	23.0	(13.2)	21.4	(10.0)	20.2	(8.3)	21.7	(11.0)
Base	*705*		*634*		*570*		*1905*	

Table 5.17 Mean Verbal fluency (animal naming) Test score

by marital status, household size and sex

	Marital status								Household Size						All	
	Married		Divorced or Separated		Single		Widowed		1		2		3 or more			
Verbal Fluency Test score																
Women	*Mean*	*(s.d.)*	*Mean*	*(s.d.)*	*Mean*	*(s.d.)*	*Mean*	*(s.d.)*	*Mean*	*(s.d.)*	*Mean*	*(s.d.)*	*Mean*	*(s.d.)*	*Mean*	*(s.d.)*
Mean score	21.3	(10.9)	22.1	(11.5)	22.6	(10.1)	20.5	(11.2)	21.4	(11.8)	21.5	(10.9)	19.8	(9.1)	21.3	(11.0)
Base		*558*		*127*		*65*		*329*		*457*		*550*		*72*		*1078*
Men																
Mean score	22.4	(11.8)	22.1	(7.6)	19.8	(7.9)	20.8	(8.4)	20.2	(8.3)	22.2	(9.8)	23.5	(17.3)	22.1	(11.1)
Base		*551*		*101*		*70*		*108*		*225*		*534*		*71*		*827*
All adults																
Mean score	21.9	(11.4)	22.1	(9.9)	21.2	(9.1)	20.6	(10.6)	21.0	(10.8)	21.9	(10.4)	21.7	(14.0)	21.7	(11.0)
Base		*1109*		*228*		*135*		*437*		*682*		*1084*		*143*		*1905*

Table 5.18 Mean Verbal fluency (animal naming) Test score

by tenure, area type and sex

	Tenure						Area type							
	Owned Outright		Owned with mortgage		Rented		Urban		Semi-rural		Rural		All	
Verbal Fluency Test score														
Women	*Mean*	*(s.d.)*	*Mean*	*(s.d.)*	*Mean*	*(s.d.)*	*Mean*	*(s.d.)*	*Mean*	*(s.d.)*	*Mean*	*(s.d.)*	*Mean*	*(s.d.)*
Mean score	21.9	(10.9)	22.4	(16.7)	19.0	(7.1)	20.9	(10.1)	21.3	(10.5)	23.9	(17.1)	21.3	(11.0)
Base		*688*		*99*		*288*		*688*		*294*		*97*		*1078*
Men														
Mean score	22.1	(8.7)	24.8	(16.5)	19.8	(12.0)	22.0	(12.7)	21.3	(7.3)	24.1	(9.8)	22.1	(11.1)
Base		*523*		*126*		*177*		*474*		*232*		*124*		*827*
All adults														
Mean score	22.0	(9.9)	23.8	(16.6)	19.3	(9.5)	21.4	(11.4)	21.3	(9.1)	24.0	(13.1)	21.7	(11.0)
Base		*1211*		*225*		*465*		*1162*		*526*		*221*		*1905*

Table 5.19 Mean Verbal fluency (animal naming) Test

by highest educational qualifications and sex, and social class and sex

	Highest educational qualifications						Social class											
	A level or above		GCSE level		No qualifications		I or II		III Non-manual		III Manual		IV		V		All	
Verbal Fluency Test score																		
Women	*Mean*	*(s.d.)*	*Mean*	*(s.d.)*	*Mean*	*(s.d.)*	*Mean*	*(s.d.)*	*Mean*	*(s.d.)*	*Mean*	*(s.d.)*	*Mean*	*(s.d.)*	*Mean*	*(s.d.)*	*Mean*	*(s.d.)*
Mean score	26.3	(14.6)	22.5	(10.3)	19.4	(9.7)	23.8	(13.0)	21.4	(9.3)	21.1	(12.3)	19.5	(12.4)	19.1	(6.3)	21.3	(11.0)
Base		*167*		*279*		*632*		*250*		*393*		*104*		*198*		*114*		*1078*
Men																		
Mean score	24.6	(9.0)	22.9	(14.6)	20.5	(9.8)	23.7	(7.3)	21.9	(12.6)	21.3	(12.8)	20.6	(8.7)	20.6	(19.1)	22.1	(11.1)
Base		*206*		*184*		*437*		*297*		*67*		*295*		*114*		*46*		*827*
All adults																		
Mean score	25.3	(11.6)	22.7	(12.4)	19.9	(9.7)	23.7	(10.0)	21.5	(9.9)	21.3	(12.7)	19.9	(11.2)	19.6	(11.7)	21.7	(11.0)
Base		*373*		*463*		*1069*		*547*		*460*		*399*		*312*		*160*		*1905*

Table 5.20 Mean Verbal Fluency (animal naming) Test score by household income and sex

	Household weekly gross income						
	Under £100	£100 but under £200	£200 but under £300	£300 but under £400	£400 but under £500	£500 or over	All
Verbal Fluency Test score							
Women	*Mean (s.d.)*	*Mean (s.d.)*	*Mean (s.d.)*	*Mean (s.d.)*	*Mean (s.d.)*	*Mean (s.d.)*	*Mean (s.d.)*
Mean score	21.3 (13.7)	19.9 (8.8)	21.2 (8.6)	21.3 (6.4)	26.4 (19.8)	24.4 (7.3)	21.3 (11.0)
Base	*171*	*384*	*190*	*88*	*41*	*62*	*1078*
Men							
Mean score	20.7 (17.3)	19.6 (7.7)	21.9 (8.3)	21.7 (7.6)	23.5 (8.8)	24.8 (10.8)	22.1 (11.1)
Base	*77*	*216*	*178*	*102*	*63*	*100*	*827*
All adults							
Mean score	21.1 (15.1)	19.8 (8.4)	21.5 (8.5)	21.5 (7.1)	24.6 (14.2)	24.7 (9.8)	21.7 (11.0)
Base	*248*	*600*	*368*	*190*	*104*	*162*	*1905*

Table 5.21 Mean Verbal fluency (animal naming) Test score by receipt of state benefits and sex

	State Benefits Received						
	Any disability related	Any income related	Any other (eg Pensions, or child related)	Household receives housing benefit	Any benefits	No benefits	All
Verbal Fluency Test score							
Women	*Mean (s.d.)*	*Mean (s.d.)*	*Mean (s.d.)*	*Mean (s.d.)*	*Mean (s.d.)*	*Mean (s.d.)*	*Mean (s.d.)*
Mean score	18.4 (6.6)	19.5 (6.4)	21.2 (11.2)	18.9 (6.8)	21.2 (11.1)	23.0 (8.1)	21.3 (11.0)
Base	*141*	*90*	*1022*	*179*	*1041*	*35*	*1078*
Men							
Mean score	19.9 (12.1)	18.2 (5.9)	21.3 (8.7)	18.6 (6.7)	21.1 (9.7)	25.3 (14.2)	22.1 (11.1)
Base	*177*	*66*	*537*	*106*	*645*	*180*	*827*
All adults							
Mean score	19.3 (10.3)	18.9 (6.2)	21.2 (10.3)	18.8 (6.8)	21.1 (10.6)	25.0 (13.4)	21.7 (11.0)
Base	*318*	*156*	*1559*	*285*	*1686*	*215*	*1905*

Table 5.22 Mean Verbal Fluency (animal naming) test score by sources of income and sex

	Sources of Regular Income							
	State benefits only		Non-state benefit income only		State benefit and other income		All	
Verbal Fluency Test score								
Women	*Mean*	*(s.d.)*	*Mean*	*(s.d.)*	*Mean*	*(s.d.)*	*Mean*	*(s.d.)*
Mean score	19.3	(9.1)	24.4	(8.0)	21.9	(11.7)	21.3	(11.0)
Base	*276*		*30*		*760*		*1078*	
Men								
Mean score	18.7	(6.7)	25.4	(14.2)	21.6	(10.2)	22.1	(11.1)
Base	*117*		*177*		*526*		*827*	
All adults								
Mean score	19.1	(8.5)	25.3	(13.6)	21.8	(11.1)	21.7	(11.1)
Base	*393*		*207*		*1286*		*1905*	

Table 5.23 Percentage of people with a TICS–m score below the cut-off for cognitive impairment, and mean TICS–m scores

by highest educational qualification and NART score (grouped)

Highest educational qualification	NART score	TICS–m total score under 21 (Cognitive impairment)	Mean TICS–m total score		Mean immediate word recall score		Mean delayed word recall score		
		Percentage	*Mean*	*(s.d.)*	*Mean*	*(s.d.)*	*Mean*	*(s.d.)*	*Base*
'A' Level or above	0–19	[4]	..	..	..	..	..	..	*20*
	20–29	9	24.0	(2.7)	4.55	(1.1)	2.77	(1.4)	*43*
	30–39	7	25.8	(3.3)	5.08	(1.5)	3.57	(1.6)	*98*
	40–50	4	26.4	(3.0)	5.24	(1.5)	3.87	(1.4)	*207*
	All	6	25.8	(3.2)	5.09	(1.4)	3.58	(1.5)	*368*
GCSE level	0–19	28	22.6	(3.8)	4.25	(1.4)	2.94	(1.6)	*74*
	20–29	17	23.8	(3.5)	4.52	(1.3)	3.05	(1.5)	*111*
	30–39	7	25.1	(2.9)	4.78	(1.3)	3.35	(1.4)	*148*
	40–50	6	25.8	(3.8)	4.88	(1.7)	3.71	(1.6)	*131*
	All	13	24.6	(3.6)	4.66	(1.4)	3.30	(1.6)	*464*
No qualifications	0–19	35	21.7	(3.8)	4.03	(1.4)	2.58	(1.4)	*433*
	20–29	18	23.5	(3.3)	4.36	(1.3)	3.02	(1.6)	*304*
	30–39	13	24.5	(3.2)	4.60	(1.4)	3.32	(1.5)	*203*
	40–50	11	25.0	(3.8)	4.70	(1.8)	3.41	(1.6)	*114*
	All	24	23.1	(3.8)	4.30	(1.4)	2.93	(1.5)	*1054*
All	0–19	33	21.9	(3.8)	4.10	(1.4)	2.64	(1.5)	*531*
	20–29	17	23.6	(3.3)	4.42	(1.3)	3.00	(1.5)	*459*
	30–39	10	25.0	(3.2)	4.77	(1.4)	3.39	(1.5)	*449*
	40–50	6	25.9	(3.5)	5.01	(1.6)	3.71	(1.5)	*453*
	All	17	24.0	(3.8)	4.55	(1.5)	3.15	(1.6)	*1892*

6 Impact of cognitive impairment

6.1 Introduction

This chapter parallels chapter 4 in examining the association between another index of mental wellbeing, this time cognitive impairment as measured by the TICS-m, and wider functioning. In chapter 4 it was demonstrated that neurotic symptoms have an impact on many aspects of day-to-day functioning over and above any effects of physical health problems. The overall aim in this chapter is to explore the extent to which cognitive impairment has an impact on functioning and on the extent of service use over and above that caused by physical health problems and neurotic symptoms. First it examines variations in mental and physical functioning, as measured by the MCS-12 and the PCS-12 respectively, with cognitive impairment. Next it looks at the association between cognitive impairment and difficulties with a range of Activities of Daily Living (ADLs), and finally it considers the association between cognitive impairment and use of a range of services.

The public health impact of dementia has been widely investigated. The impact of more broad categories of cognitive impairment has received less research. Several community studies have found an association between impaired cognitive function and higher subsequent mortality rates. This has been found to be independent of age and physical health status (Liu *et al*, 1990; Gale *et al*, 1996) and particularly strong for death from stroke (Gale *et al*, 1996). Lower cognitive function has also been found to predict incidence of first stroke (Ferucci *et al*, 1996), and onset of dementia after a ten year period (Elias *et al*, 2000). As well as its prognostic significance, cognitive impairment has been found to be associated with increased disablement. In a US community study cognitive impairment was associated with lower instrumental activities of daily living scores, but not with physical health status (Fillenbaum *et al*, 1988).

The impact of cognitive impairment on service use remains controversial. In the Medical Research Council Cognitive Function and Ageing (CFA) Study of UK people aged 65 and over, cognitive impairment was strongly associated with residence in an institutional care facility, particularly if combined with physical disability. However people with cognitive disability alone were less likely to receive personal or household care, meals on wheels or home help than those with physical disability alone (Melzer *et al*, 1999). Reduced use of domiciliary services was also found in a UK sample of women aged 75 and over, comparing those with cognitive impairment to those with physical impairment and comparable dependency (Ely *et al*, 1997). CFA Study participants with cognitive disability alone were also less likely to have acute or outpatient hospital contact than those with physical disability alone (Melzer *et al*, 1999). A US study on the other hand found that cognitive impairment strongly predicted hospitalisation over a one-year follow-up period after adjustment for activities of daily living (Weiler *et al*, 1991).

6.2 SF-12 mental and physical health component summary scores

Mean mental (MCS-12) and physical (PCS-12) health component summary scores were lower, indicating poorer functioning, among both men and women with cognitive impairment than among those with higher TICS-m scores. The difference in mean MCS-12 scores between those with and without evidence of cognitive impairment was statistically significant in women (where the mean score was 49.7 among the cognitively impaired compared with 51.6 among those with TICS-m scores of 21 or more), but not in men. The difference in PCS-12 scores between those with and without cognitive impairment was statistically significant in both sexes. Across all respondents, the mean PCS-12 score was 42.3 among those with a TICS-m score below 21 and 45.7 among those with a TICS-m score of 21 or more. *Table 6.1*

As noted in Chapter 4, mean MCS-12 and PCS-12 scores decreased significantly with increasing levels of long-term physical health problems. At almost every level of physical health problems, mean

MCS-12 and PCS-12 scores were slightly lower among respondents with evidence of cognitive impairment than among those with TICS-m scores of 21 or more, most differences were not statistically significant. *Table 6.2*

MCS-12 and PCS-12 scores were lower among people with CIS-R scores of 12 or more (indicative of clinically significant levels of neurotic symptoms), than among those with CIS-R scores below 12 (see chapter 4). Within these two groups based on CIS-R score, mean MCS-12 and PCS-12 scores were slightly lower among respondents with cognitive impairment than among those without, but the difference was only statistically significant for mean PCS-12 score among respondents with CIS-R scores below 12. *Table 6.3*

The analyses above suggest that cognitive impairment, as indicated by a TICS-m score below 21, although associated with poorer mental and physical functioning, has only a limited impact over and above the effect of any underlying poor physical health and common mental health problems.

6.3 Difficulties with activities of daily living (ADLs)

6.3.1 Number of ADLs

People with evidence of cognitive impairment were more likely than those without to have difficulty with at least one activity of daily living, (52% compared with 34%). This difference between respondents with lower and higher TICS-m scores was observed at all levels of physical health problems, but was less marked among those with more physical health problems among whom the majority reported difficulties with ADLs anyway. The proportion of people having difficulty with one or more ADLs was significantly higher among people with cognitive impairment compared with those with TICS-m scores of 21 or more in the group with no physical health problems, (32% compared with 12%), and those with just one organ system affected, (53% and 34%, respectively). Similarly, the difference in likelihood of having difficulty with at least one ADL between those with and without cognitive impairment was evident among respondents with and without mental health problems, but was only statistically significant among people with CIS-R scores below 12.
Tables 6.4 and 6.5 and Figure 6.1

Figure 6.1 Proportion of people with one or more ADL difficulties by cognitive function (TICS-m score) at different levels of physical health problems

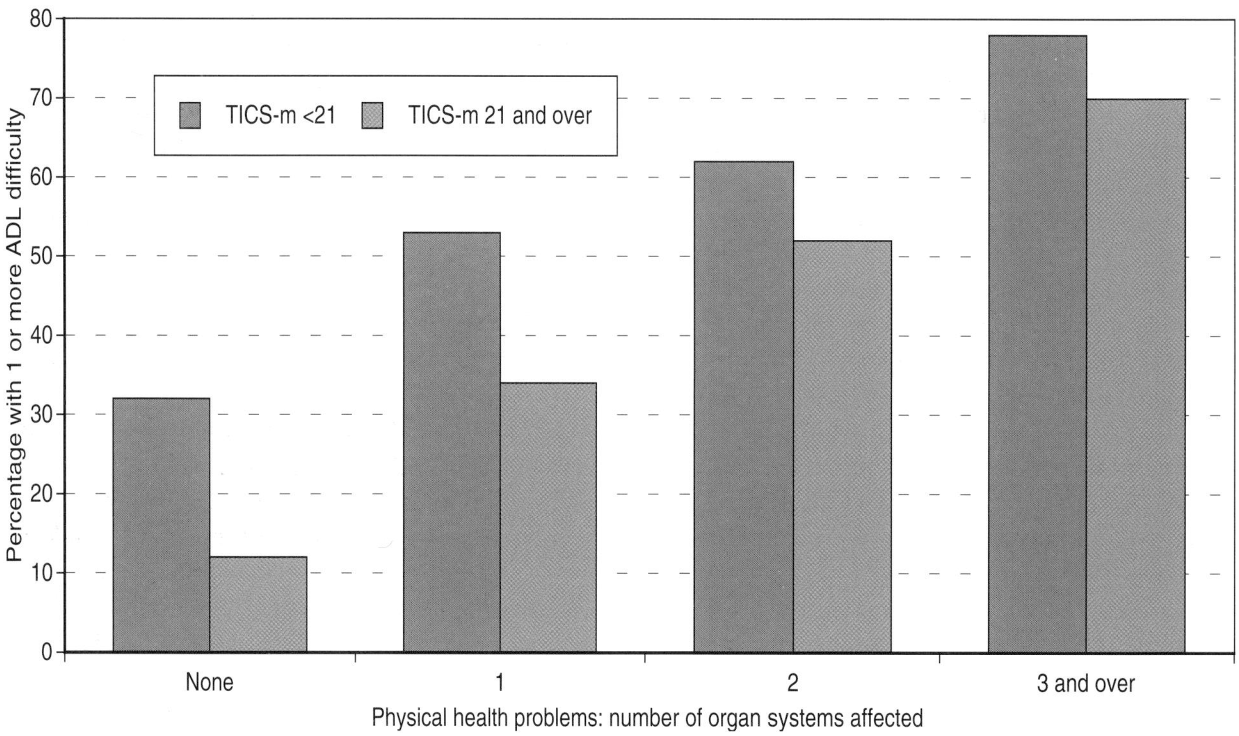

6.3.2 Type of ADL difficulty

People with signs of cognitive impairment were significantly more likely to have difficulty with each of the seven ADLs asked about, except medical care, than were those with a TICS-m score of 21 or more. Paperwork was the one of the ADLs that people most frequently reported difficulty with, reported by 14% of respondents. People with cognitive impairment were three times as likely as those without to experience difficulty with paperwork, (30% compared with 10%). Difficulties with managing money were reported by a very small percentage of respondents (3%), but people with cognitive impairment were four times as likely as those without to report difficulty (8% compared with 2%). People with cognitive impairment were also twice as likely as those without to have difficulties with using transport, (22% compared with 11%) and personal care, (14% compared with 7%). They were about one and half times as likely as people without cognitive impairment to have difficulty with household activities (18% compared with 12%), and practical activities (35% compared with 27%). *Table 6.4*

The extent of the impact of cognitive impairment on performance of everyday activities varied with level of physical health problems as well as between types of activity. The ADL that was most consistently (across different levels of physical health) associated with greater likelihood of difficulty among respondents with cognitive impairment than among those without was paperwork. Among people with no longstanding physical health problems, difficulty with paperwork was 4 times as likely among people with cognitive impairment than among those without (25% compared with 6%). Among those with one and two organ systems affected difficulty with paperwork was about three times as likely and among those with three organ systems affected they were about twice as likely if respondents were cognitively impaired. *Table 6.4*

At both levels of CIS-R scores, (below 12 and 12 or over), respondents with TICS-m scores below 21 were more likely than those with higher TICS-m scores to report having difficulties with each of the seven ADLs considered. However, the magnitude of differences between people with and without cognitive impairment varied with level of CIS-R scores and between types of activity. The activities on which cognitive impairment had the greatest impact over and above the effect of common mental disorder were personal care, managing money and dealing with paperwork. *Table 6.5*

These results suggest that cognitive impairment is associated with difficulties with activities of daily living over and above effects of any underlying physical health problems and neurotic symptoms. As would be expected, the relative impact of cognitive impairment, physical health problems, and neurotic symptoms varies between different types of activities.

6.4 Impact of cognitive impairment on use of services

6.4.1 Community based service use in the year prior to interview

The likelihood of GP consultation both for physical and for mental complaints was the same among people with and without evidence of cognitive impairment. Almost three-quarters of respondents (72%) in each case had consulted their GP about a physical complaint in the year prior to interview, and 8% in each case had consulted about a mental complaint. Similarly, there was no difference between people with and without cognitive impairment in the likelihood of having received any other community care in the year prior to interview. Just 6% of respondents in each case had received any such care. As noted in chapter 4, the likelihood of use of specific types of community service was very low, and there were no differences between respondents with and without cognitive impairment. *Tables 6.6 and 6.7*

6.4.2 Hospital service use in the quarter prior to interview

People with evidence of cognitive impairment were no more likely than those without to have had hospital inpatient or outpatient treatment, for mental or physical complaints, in the quarter prior to the interview. However, as noted in chapter 4, the level of use of these services was generally low. *Table 6.7*

6.4.3 *Assessment of the impact of cognitive impairment on service use*

As mentioned above, cognitive impairment was not associated with differential use of any of the services examined in this report. This applied irrespective of CIS-R score level, and, across different levels of physical health problems.

Tables 6.6 and 6.8

Since cognitive impairment was associated with increased levels of difficulties with activities of daily living and, to a lesser extent, with poorer mental and physical functioning, the results above suggest a possible deficit of help-seeking or of access to services among people with cognitive impairment. However, table 6.9, illustrates that there was no consistent association between cognitive impairment and service use even when the extent of problems with activities of daily living were taken into account. Of the types of service examined, it might be expected that level of difficulties with activities of daily living would be most likely to impact on use of community based care. People with evidence of cognitive impairment do appear to be less likely than those without to access these services at each level of ADL difficulties, but these differences are not statistically significant. *Table 6.9 and Figure 6.2*

References

Elias M F, Beiser A, Wolf P A, Au R, White R F and D'Agostino R B (2000) The preclinical phase of Alzheimer disease: a 22-year prospective study of the Framingham cohort. *Archives of Neurology* **57**, 808–813.

Ely M, Brayne C, Huppert F A, O'Connor D W and Pollitt P A (1997) Cognitive impairment: a challenge for community care. A comparison of the domiciliary service receipt of cognitively impaired and equally dependent physically impaired elderly women. *Age and Ageing* **26**, 301–308.

Ferucci L, Guralnik J M, Salive M E, Pahor M, Corti M, Baroni A and Havlik R J (1996) Cognitive impairment and risk of stroke in the older population. *Journal of the American Geriatrics Society* **44**, 237–241.

Fillenbaum G G, Hughes D C, Heyman A, George L K and Blazer D G (1988) Relationship of health and demographic characteristics to Mini-Mental State Examination score among community residents. *Psychological Medicine* **18**, 719–726.

Gale C R, Martyn C N and Cooper C (1996) Cognitive impairment and mortality in a cohort of elderly people. *British Medical Journal* **312**, 608–611.

Liu I Y, Lacroix A Z, White L R, Kittner S J and Wolf P A (1990) Cognitive impairment and mortality: a study of possible confounders. *American Journal of Epidemiology* **132**, 136–143.

Melzer D, McWilliams B, Brayne C, Johnson T and Bond J (1999) Profile of disability in elderly people: estimates from a longitudinal population study. *British Medical Journal* **318**, 1108–1111.

Weiler P G, Lubben J E and Chi I (1991) Cognitive impairment and hospital use. *American Journal of Public Health* **81**, 1153–1157.

Figure 6.2 Proportion of people receiving community care services by cognitive function (TICS-m score) among people with different numbers of ADL difficulties

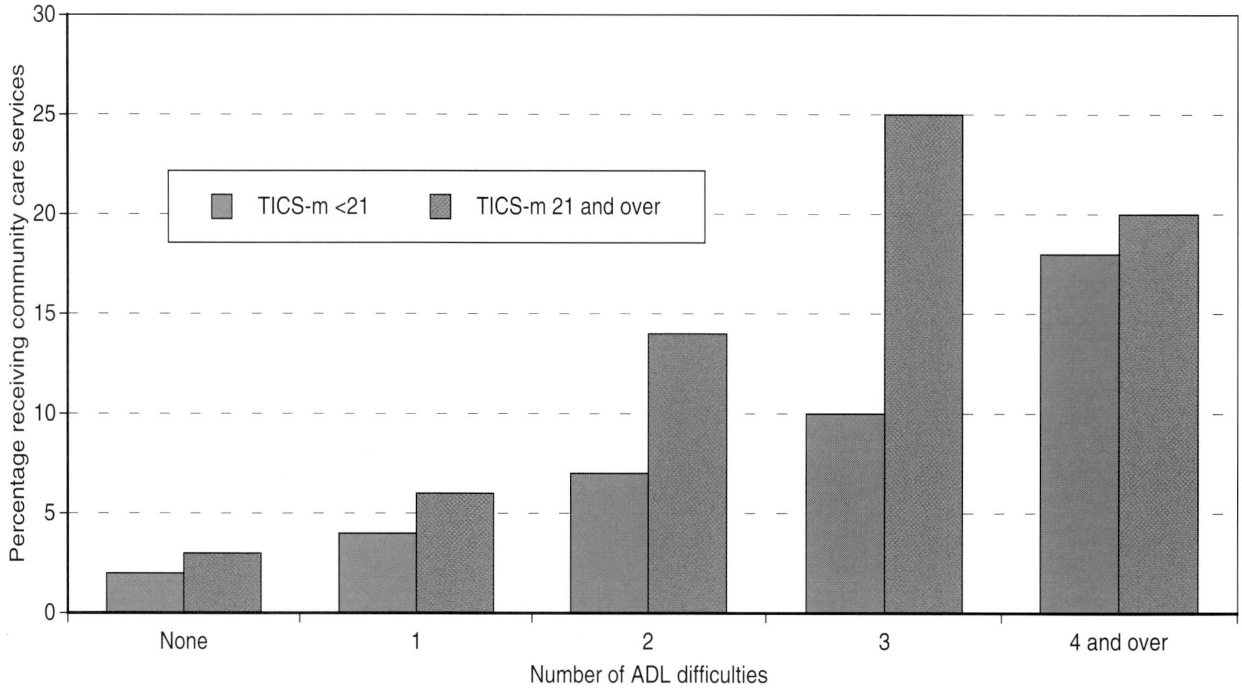

Table 6.1 Mean SF-12 mental and physical health summary scores by TICS-m score (grouped) and sex

	SF-12 Mental Health Summary Score								
	Men		Women		All		*Men*	*Women*	*All*
	Mean	(s.d.)	Mean	(s.d.)	Mean	(s.d.)	*Base*	*Base*	*Base*
TICS-m score									
Under 21 (Cognitive impairment)	52.1	(8.8)	49.7	(11.4)	50.8	(10.4)	*157*	*227*	*384*
21 and over	53.3	(8.3)	51.6	(9.1)	52.4	(8.8)	*703*	*902*	*1605*
All	53.1	(8.4)	51.3	(9.6)	52.1	(9.1)	*860*	*1129*	*1989*
	SF-12 Physical Health Summary Score								
TICS-m score									
Under 21 (Cognitive impairment)	43.1	(12.2)	41.7	(12.7)	42.3	(12.5)	*157*	*227*	*384*
21 and over	45.9	(11.9)	45.6	(12.2)	45.7	(12.1)	*703*	*902*	*1605*
All	45.4	(12.0)	44.8	(12.4)	45.1	(12.2)	*860*	*1129*	*1989*

Table 6.2 Mean SF-12 mental and physical health summary scores by TICS-m score (grouped) and extent of longstanding physical health problems

		MCS-12		PCS-12		
		Mean	(s.d.)	Mean	(s.d.)	*Base*
Number of organ systems affected by longstanding physical health problems	**TICS-m score**					
None	Under 21 (Cognitive impairment)	54.6	(8.0)	52.3	(5.6)	*108*
	21 and over	54.2	(7.3)	53.2	(5.7)	*564*
	All	54.3	(7.4)	53.0	(5.7)	*672*
One	Under 21 (Cognitive impairment)	50.3	(10.3)	40.8	(12.0)	*157*
	21 and over	52.5	(8.7)	45.5	(11.2)	*546*
	All	52.0	(9.1)	44.4	(11.5)	*703*
Two	Under 21 (Cognitive impairment)	50.4	(11.2)	36.7	(11.3)	*68*
	21 and over	50.8	(9.7)	40.6	(12.0)	*328*
	All	50.7	(10.0)	40.0	(12.0)	*396*
Three or more	Under 21 (Cognitive impairment)	44.3	(11.0)	31.5	(11.0)	*50*
	21 and over	49.3	(10.2)	31.3	(12.1)	*165*
	All	48.2	(10.5)	31.4	(11.9)	*215*
All	Under 21 (Cognitive impairment)	50.8	(10.4)	42.3	(12.5)	*384*
	21 and over	52.4	(8.8)	45.7	(12.1)	*1605*
	All	52.1	(9.1)	45.1	(12.2)	*1989*

Table 6.3 Mean SF-12 mental and physical health summary scores

by TICS-m score (grouped) and CIS-R score (grouped)

		MCS-12		PCS-12		
		Mean	(s.d.)	Mean	(s.d.)	*Base*
CIS-R score	**TICS-m score**					
Under 21	Under 21 (Cognitive impairment)	53.3	(8.1)	44.1	(11.8)	*321*
	21 and over	53.7	(7.5)	46.9	(11.3)	*1452*
	All	53.6	(7.6)	46.4	(11.5)	*1773*
12 and over	Under 21 (Cognitive impairment)	37.1	(10.8)	32.8	(11.7)	*63*
	21 and over	39.4	(10.1)	33.8	(12.9)	*153*
	All	38.7	(10.4)	33.5	(12.5)	*216*
All	Under 21 (Cognitive impairment)	50.8	(10.4)	42.3	(12.5)	*384*
	21 and over	52.4	(8.8)	45.7	(12.1)	*1605*
	All	52.1	(9.1)	45.1	(12.2)	*1989*

Table 6.4 Difficulties with ADLs

by TICS-m score (grouped) and extent of longstanding physical health problems

	Number of organ systems affected by longstanding physical health problems										
	0		1		2		3 and over		All		
	TICS-m score										
	Under 21	21 and over	Under 21	21 and over	Under 21	21 and over	Under 21	21 and over	Under 21	21 and over	All
Number of ADL difficulties	%	%	%	%	%	%	%	%	%	%	%
0	68	88	47	66	38	48	22	30	48	66	63
1	24	8	17	20	21	25	22	20	20	17	18
2	4	2	10	5	20	10	6	13	9	6	7
3	3	0	9	4	9	10	18	18	8	5	6
4 or more	1	1	17	5	13	8	32	19	14	6	7
Any ADL difficulty	**32**	**12**	**53**	**34**	**62**	**52**	**78**	**70**	**52**	**34**	**37**
	Percentage reporting each difficulty										
Type of ADL difficulty											
Practical activities	10	7	40	26	43	47	62	63	35	27	29
Dealing with paperwork	25	6	33	11	29	12	36	16	30	10	14
Household activities	1	2	20	10	23	19	42	36	18	12	13
Using transport	3	1	28	10	24	17	43	35	22	11	13
Managing money	4	1	9	2	5	2	21	4	8	2	3
Personal care	-	0	14	5	20	12	39	27	14	7	9
Medical care	-	-	2	2	1	1	18	3	3	1	2
Base	*107*	*559*	*155*	*544*	*67*	*328*	*50*	*165*	*380*	*1598*	*1978*

Table 6.5 Difficulties with ADLs

by TICS-m score (grouped) and CIS-R score (grouped)

	CIS-R score						
	Under 12		12 and over		All		
	TICS-m score						
	Under 21	21 and over	Under 21	21 and over	Under 21	21 and over	All
Number of ADL difficulties	%	%	%	%	%	%	%
0	54	70	20	30	48	66	63
1	22	17	11	20	20	17	18
2	9	6	11	10	9	6	7
3	7	4	17	16	8	5	6
4 or more	8	4	41	25	14	6	7
Any ADL difficulty	**46**	**30**	**80**	**70**	**52**	**34**	**37**
	Percentage reporting each difficulty						
Type of ADL difficulty							
Practical activities	29	24	67	61	35	27	29
Dealing with paperwork	27	8	45	29	30	10	14
Household activities	13	9	48	42	18	12	13
Using transport	15	9	59	35	22	11	13
Managing money	6	1	21	5	8	2	3
Personal care	9	6	3:	24	14	7	9
Medical care	2	1	8	5	3	1	2
Base	*318*	*1447*	*62*	*153*	*380*	*1600*	*1980*

Table 6.6 Use of community services

by TICS-m scores (grouped)

	TICS-m score		
	Under 21	21 and over	All
	Percentage using each service		
Community service use in last year			
Seen psychiatrist in last year	1	0	0
Seen psychologist in last year	-	0	0
Seen community psychiatric nurse in last year	1	0	1
Seen community learning difficulty nurse in last year	-	-	-
Used other nursing services in last year	3	4	4
Seen social worker in last year	1	1	1
Used self help/support group in last year	-	1	1
Used home help/home care worker in last year	1	1	1
Seen outreach worker in last year	-	0	0
Any community care in last year	**6**	**6**	**6**
Base	*384*	*1605*	*1989*

Table 6.7 Use of services

by TICS-m scores (grouped) and the extent of longstanding physical health problems

	Number of organ systems affected by longstanding physical health problems										
	0		1		2		3 and over		All		
	TICS-m score										
	Under 21	21 and over	Under 21	21 and over	Under 21	21 and over	Under 21	21 and over	Under 21	21 and over	All
	Percentage using each service										
Service use											
Visited GP with physical complaint in last year	46	57	77	73	84	84	95	93	72	72	72
Visited GP with mental complaint in last year	5	7	7	6	9	10	14	12	8	8	8
Any GP visit in last year	**50**	**60**	**78**	**74**	**85**	**85**	**95**	**94**	**74**	**73**	**73**
Had outpatient visit for physical complaint in last quarter	11	14	28	25	28	30	53	49	26	24	25
Had outpatient visit for mental complaint in last quarter	0	1	2	-	-	0	-	-	1	0	0
Any outpatient visit in last quarter	**11**	**14**	**29**	**25**	**28**	**30**	**53**	**49**	**27**	**25**	**25**
Had inpatient stay for physical complaint in last quarter	2	2	2	6	7	4	9	9	4	4	4
Had inpatient stay for mental complaint in last quarter	-	-	-	0	-	-	-	-	-	0	0
Any inpatient stay in last quarter	**2**	**2**	**2**	**6**	**7**	**4**	**9**	**9**	**4**	**4**	**4**
Any community care in last year	**2**	**3**	**7**	**6**	**5**	**8**	**13**	**14**	**6**	**6**	**6**
Base	*108*	*564*	*157*	*546*	*68*	*328*	*50*	*165*	*384*	*1605*	*1989*

Table 6.8 Use of services

by TICS-m score (grouped) and CIS-R score (grouped)

	CIS-R score						
	Under 12		12 and over		All		
	TICS-m score						
	Under 21	21 and over	Under 21	21 and over	Under 21	21 and over	All
	Percentage using each service						
Service use							
Visited GP with physical complaint in last quarter	69	70	89	89	72	72	72
Visited GP with mental complaint in last year	9	10	5	5	9	8	8
Any GP visit in last year	**72**	**71**	**71**	**75**	**77**	**74**	**73**
Had outpatient visit for physical complaint in last quarter	23	23	41	43	26	24	25
Had outpatient visit for mental complaint in last quarter	1	0	1	1	1	0	0
Any outpatient visit in last quarter	**24**	**23**	**42**	**44**	**27**	**25**	**25**
Had inpatient stay for physical complaint in last quarter	3	4	10	7	4	4	4
Had inpatient stay for mental complaint in last quarter	-	-	-	1	-	0	0
Any inpatient stay in last quarter	**3**	**4**	**10**	**8**	**4**	**4**	**4**
Any community care in last year	**4**	**5**	**14**	**15**	**6**	**6**	**6**
Base	*321*	*1452*	*63*	*153*	*384*	*1605*	*1989*

Table 6.9 Use of services

by TICS-m scores (grouped) and number of ADL difficulties

	Number of ADL difficulties												
	0		1		2		3		4+		All		
	TICS-m score												
	Under 21	21 and over	Under 21	21 and over	Under 21	21 and over	Under 21	21 and over	Under 21	21 and over	Under 21	21 and over	All
	Percentage using each service												
Service use													
Visited GP with physical complaint in last quarter	62	66	73	78	84	83	90	91	90	92	72	72	72
Visited GP with mental complaint in last year	4	6	5	8	12	9	9	13	20	21	8	8	8
Any GP visit in last year	**63**	**68**	**75**	**81**	**84**	**85**	**92**	**91**	**93**	**93**	**74**	**73**	**73**
Had outpatient visit for physical complaint in last quarter	17	20	25	25	16	38	55	47	53	40	26	24	25
Had outpatient visit for mental complaint in last quarter	-	0	1	-	2	1	-	-	3	1	1	0	0
Any outpatient visit in last quarter	**17**	**20**	**26**	**25**	**18**	**38**	**55**	**47**	**54**	**40**	**27**	**25**	**25**
Had inpatient stay for physical complaint in last quarter	1	3	5	8	9	5	9	9	7	9	4	4	4
Had inpatient stay for mental complaint in last quarter	-	-	-	0	-	-	-	-	-	-	-	0	0
Any inpatient stay in last quarter	**1**	**3**	**5**	**8**	**9**	**5**	**9**	**9**	**7**	**9**	**4**	**4**	**4**
Any community care in last year	**2**	**3**	**4**	**6**	**7**	**14**	**10**	**25**	**18**	**20**	**6**	**6**	**6**
Base	*177*	*1033*	*78*	*283*	*45*	*104*	*31*	*87*	*49*	*90*	*384*	*1604*	*1988*

7 Comorbidity

7.1 Introduction

Cognitive impairment has been frequently found to be associated with depression in older people (Fuhrer *et al*, 1992). Associations with increased disablement have been described in the previous chapter. Cognitive impairment is recognised to be associated with specific physical disorders including brain disorders (for example Parkinson's disease and stroke), and risk factors for vascular disease (for example hypertension and diabetes). Associations with global measures of physical health in community populations have not been established (Fillenbaum *et al*, 1988).

This chapter reports on the variation in likelihood of having possible cognitive impairment, as measured by the TICS-m, with level of neurotic symptoms and prevalence of neurotic disorders, as measured by the CIS-R, and long-standing physical health problems.

7.2 Distribution of CIS-R scores, and prevalence of neurotic disorders by TICS-m scores and sex

Both men and women with TICS-m scores indicative of possible cognitive impairment were more likely than those with higher TICS-m scores to have CIS-R scores indicative of clinically significant levels of neurotic symptoms. About one out of six (16%) of all respondents with a TICS-m score below 21, compared with under one out of ten (9%) of those with a TICS-m score of 21 or more, had a CIS-R score of 12 or over. However, the difference between those with TICS-m scores above and below the threshold reached statistical significance only in men. *Table 7.1*

Similarly, both men and women with cognitive impairment were more likely than those without to have some neurotic disorder, (18% compared with 10% respectively had some neurotic disorder). *Table 7.2*

Figure 7.1 Distribution of physical health problems by level of TICS-m score and sex

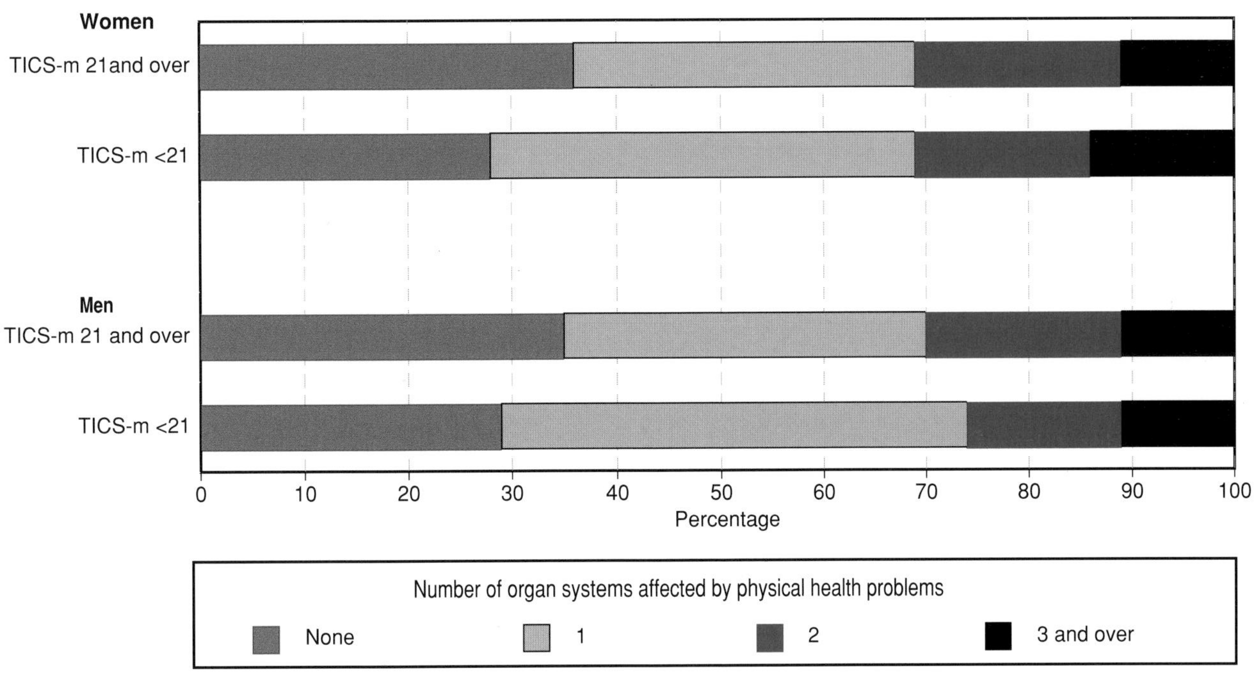

7.3 Prevalence of physical health problems by TICS-m scores and sex

Respondents of both sexes with a TICS-m score below 21 were more likely than those with higher scores to have a longstanding physical health problems, but this difference was statistically significant only among women.

Table 7.3 and Figure 7.1

The results discussed above support the idea that cognitive impairment is associated with increased likelihood of having common mental disorders and physical health problems.

References

Fillenbaum G G, Hughes D C, Heyman A, George L K and Blazer D G (1988) Relationship of health and demographic characteristics to Mini-Mental State Examination score among community residents. *Psychological Medicine* **18**, 719–726.

Fuhrer R, Antonucci T C, Gagnon M, Dartigues J-F and Barberger-Gateau P (1992) Depressive symptomatology and cognitive functioning: an epidemiological survey in an elderly community sample in France. *Psychological Medicine* **22**, 159–172.

Table 7.1 Distribution of CIS-R scores (grouped) by TICS-m scores (grouped) and sex

	TICS-m score		
	Under 21	21 and over	All
CIS-R score	%	%	%
Women			
0–5	69	73	72
6–11	15	16	16
Under 12	**83**	**89**	**88**
12–17	9	6	7
18 and over	8	5	5
12 and over	**17**	**11**	**12**
Base	*227*	*902*	*1129*
Men			
0–5	73	82	80
6–11	13	11	12
Under 12	**86**	**93**	**92**
12–17	10	3	4
18 and over	5	4	4
12 and over	**14**	**7**	**8**
Base	*157*	*703*	*860*
All adults			
0–5	70	77	76
6–11	14	14	14
Under 12	**84**	**91**	**90**
12–17	9	5	6
18 and over	7	4	5
12 and over	**16**	**9**	**10**
Base	*384*	*1605*	*1989*

Table 7.2 Prevalence of neurotic disorder by TICS-m scores (grouped) and sex

	TICS-m score		
	Under 21	21 and over	All
Neurotic disorder	%	%	%
Women			
Mixed anxiety and depressive disorder	11	7	8
Generalised anxiety disorder	4	3	4
Depressive episode	3	1	1
All phobias	1	1	1
Obsessive compulsive disorder	2	1	1
Panic disorder		1	0
Any anxiety disorder	6	5	5
Any neurotic disorder excluding mixed anxiety and depression	8	5	6
Any neurotic disorder	**19**	**13**	**14**
Base	*227*	*902*	*1129*
Men			
Mixed anxiety and depressive disorder	10	4	5
Generalised anxiety disorder	2	3	2
Depressive episode	2	1	2
All phobias	2	0	1
Obsessive compulsive disorder	1	0	0
Panic disorder	0	0	0
Any anxiety disorder	4	3	3
Any neurotic disorder excluding mixed anxiety and depression	6	4	4
Any neurotic disorder	**16**	**8**	**9**
Base	*157*	*703*	*860*
All Adults			
Mixed anxiety and depressive disorder	10	6	7
Generalised anxiety disorder	3	3	3
Depressive episode	3	1	1
All phobias	2	1	1
Obsessive compulsive disorder	1	0	1
Panic disorder	0	0	0
Any anxiety disorder	5	4	4
Any neurotic disorder excluding mixed anxiety and depression	7	4	5
Any neurotic disorder	**18**	**10**	**12**
Base	*384*	*1605*	*1989*

Table 7.3 Prevalence of longstanding physical health problems

by TICS-m scores (grouped) and sex

	TICS-m score		
	Under 21	21 and over	All
	%	%	%
Women			
Has longstanding physical illness			
Yes	72	64	66
No	28	36	34
Number of organ systems affected by longstanding physical illness			
None	28	36	34
1	41	33	34
2	17	21	20
3 or more	14	11	11
Base	*227*	*901*	*1128*
Men			
Has longstanding physical illness			
Yes	71	65	66
No	29	35	34
Number of organ systems affected by longstanding physical illness			
None	29	35	34
1	45	35	37
2	16	19	18
3 or more	11	11	11
Base	*156*	*702*	*858*
All			
Has longstanding physical illness			
Yes	72	65	66
No	28	35	34
Base	*383*	*1603*	*1986*
Number of organ systems affected by longstanding physical illness			
None	28	35	34
1	43	34	36
2	17	20	19
3 or more	12	11	11
Base	*383*	*1603*	*1986*

A

Appendix A Statistical terms and their interpretation

A1 Confidence intervals

The percentages and means quoted in the text of this report represent summary information about a variable (eg CIS-R score) based on the sample of people interviewed in this study. However, extrapolation from these sample statistics is required in order to make inferences about distribution of that particular variable in the population. This is done by calculating confidence intervals around the statistic in question. These confidence intervals indicate the range within which the 'true' (or population) percentage is likely to lie. Where 95% confidence intervals are calculated, this simply indicates that one is '95% confident' that the population percentage lies within this range. (More accurately, it indicates that, if repeated samples were drawn from the population, the percentage would lie within this range in 95% of the samples.)

Confidence intervals are calculated on the basis of the sampling error (see below). The upper 95% confidence intervals are calculated by adding the sampling error (SE) multiplied by 1.96 to the sample percentage or mean. The lower confidence interval is derived by subtracting the same value. 99% confidence intervals can also be calculated, by replacing the value 1.96 by the value 2.58.

A2 Sampling errors

The sampling error is a measure of the degree to which a percentage (or other summary statistic) would vary if repeatedly calculated in a series of samples. It is used in the calculation of confidence intervals and statistical significance tests. In this survey simple random sampling did not take place, a multi-stage stratified sampling design was used. In addition, the data was weighted firstly to take account of differing selection probabilities and, secondly, to compensate for non-response using post-stratification. To take account of the complex sample design and weighting procedures used in this survey, sampling errors were calculated using STATA and the sampling errors associated with key estimates are shown in Tables A.1–A.4 below. However, this does not affect the interpretation of the sampling errors or their use in the calculation of confidence intervals.

The effect of a complex sampling design on the precision of survey estimates is usually quantified by means of the design factor (deft). For any survey estimate, the deft is calculated as the ratio of the standard error allowing for the full complexity of the survey design to the standard error assuming a simple random sample. The standard error based on a simple random sample multiplied by the deft gives the standard error of a complex design.

$$se(p) = deft \times se(p)_{sys}$$

where:

$$se(p)_{sys} = \sqrt{\frac{p(1-p)}{N}}$$

The formula to measure whether the differences between the percentages is likely to be due entirely to sampling error for a complex design is:

$$se(p_1\text{-}p_2) = \sqrt{\frac{deft_1^{\,2} * p_1(100-p_1)}{n_1} + \frac{deft_2^{\,2} * p_2(100-p_2)}{n_2}}$$

where p_1 and p_2 are observed percentages for the two sub-samples and n_1 and n_2 are the sub-sample sizes. The 95% confidence interval for the difference between two percentages is then given by;

$$(p_1-p_2) \text{ +/- } 1.96 \times se(p_1\text{-}p_2)$$

If this confidence interval includes zero then the observed difference is considered to be a result of chance variation in the sample. If the interval does not include zero then it is unlikely (less than 5% probability) that the observed differences could have occurred by chance. The standard errors of survey measures which are not presented in the following tables for sample subgroups may be estimated by applying an appropriate value of deft to the sampling error. The choice of an appropriate value of deft will vary according to whether the basic survey measure is included in the tables. Since most deft values are relatively small (1.1 or less) the absolute effect of adjusting sampling errors to take account of the survey's complex design will be small. In most cases it will result in an increase of less than 10% over the standard error assuming simple random sampling. However, for some regional estimates the deft is greater and caution should be exercised when considering the significance of apparent differences between regions. Whether it is considered necessary to use deft or to use the basic estimates of standard errors assuming a simple random sample is a matter of judgement and depends chiefly on how the survey results will be used.

Sampling errors have been calculated for other estimates and have been used to test the statistical significance of differences for this report. In general only statistically significant differences are commented on in the report unless specifically stated otherwise. Tables of additional sampling errors can be found in the Technical Report of the survey.

A3 Significance

It is stated in the text of the report that some differences are 'significant'. This indicates that it is unlikely that a difference of this magnitude would be found due to chance alone. Specifically, the likelihood that the difference would occur simply by chance is less than 5%. This is conventionally assumed to be in frequent enough to discount chance as an explanation for the finding.

A4 Multiple Logistic Regression (MLR) and Odds Ratios (OR)

Logistic regression analysis has been used in the analysis of the survey data to provide a measure of the effect of, for example, various characteristics on the levels of neurotic disorder in carers. Unlike the cross tabulations presented elsewhere in the report, MLR estimates the effect of any variable while controlling for the confounding effect of other variables in the analysis.

Logistic regression produces an estimate of the probability of an event occurring when an individual is in a particular category compared to a reference category. For example, to provide an estimate of the likelihood that a person will have a CIS-R score of 12 or more if they are divorced or separated compared with the likelihood if they are married. This effect is measured in terms of odds. The amount by which the odds are actually increased is shown by the Adjusted Odds Ratio (OR). To determine whether this increase is due to chance rather than to the effect of the variable one must consult the associated 95% confidence interval. In tables showing adjusted odds ratios, ORs which are statistically significant at the 95% level are indicated by a single asterisk * and ORs which are statistically significant at the 99% level are indicated by double asterisks **.

A forward stepwise variable entry method was used in the MLR analyses presented in this report.

Confidence intervals around an Odds Ratio

If the confidence interval does not include 1.00 then the OR is likely to be significant – that is, the association between the variable and the odds of high levels of neurotic symptoms is unlikely to be due to chance. If the interval includes 1.00, then it is possible that the 'true' OR is actually 1.00, that is no increase in odds can be attributed to the variable.

Odds Ratios and how to use them multiplicatively

The odds ratios presented in the tables show the adjusted odds due solely to membership of one particular category – for example being divorced or separated rather than married. However odds for more than one category can be combined by multiplying them together.

Table A.1 Standard errors and 95% confidence intervals for the distribution of CIS-R scores by age (Table 2.1)

Base	Characteristic	% (p) (adj)	Sample size	True standard error of p	Deft	95% confidence interval LL	UL
Women							
60- to 64-year-olds	0–5	71.27	407	2.31	1.03	66.74	75.79
	6–11	15.51	407	1.88	1.05	11.82	19.21
	Under 12	86.78	407	1.70	1.01	83.44	90.12
	12–17	7.11	407	1.32	1.03	4.52	9.70
	18 and over	6.11	407	1.15	0.96	3.86	8.35
	12 and over	13.22	407	1.70	1.01	9.88	16.56
65- to 69-year-olds	0–5	73.10	373	2.21	0.96	68.77	77.44
	6–11	14.69	373	1.73	0.94	11.29	18.08
	Under 12	87.79	373	1.87	1.10	84.13	91.45
	12–17	6.68	373	1.48	1.15	3.78	9.59
	18 and over	5.53	373	1.28	1.08	3.02	8.04
	12 and over	12.21	373	1.87	1.10	8.55	15.87
70- to 74-year-olds	0–5	72.02	356	2.25	0.94	67.60	76.43
	6–11	17.42	356	1.95	0.97	13.60	21.25
	Under 12	89.44	356	1.66	1.02	86.19	92.69
	12–17	6.61	356	1.46	1.11	3.75	9.48
	18 and over	3.94	356	0.90	0.87	2.18	5.71
	12 and over	10.56	356	1.66	1.02	7.31	13.81
All (60- to 74-year olds)	0–5	72.13	1136	1.31	0.99	69.56	74.71
	6–11	15.73	1136	1.12	1.04	13.53	17.93
	Under 12	87.86	1136	1.00	1.03	85.90	89.82
	12–17	6.82	1136	0.84	1.12	5.18	8.46
	18 and over	5.31	1136	0.65	0.98	4.04	6.59
	12 and over	12.14	1136	1.00	1.03	10.18	14.10
Men							
60- to 64-year olds	0–5	74.00	332	2.70	1.12	68.71	79.29
	6–11	13.02	332	2.13	1.15	8.85	17.20
	Under 12	87.02	332	1.89	1.02	83.33	90.72
	12–17	5.72	332	1.40	1.10	2.98	8.46
	18 and over	7.26	332	1.50	1.05	4.32	10.19
	12 and over	12.98	332	1.89	1.02	9.28	16.67
65- to 69-year olds	0–5	83.64	295	2.18	1.01	79.37	87.91
	6–11	11.68	295	2.03	1.09	7.69	15.66
	Under 12	95.32	295	1.11	0.90	93.14	97.49
	12–17	3.16	295	0.91	0.89	1.38	4.94
	18 and over	1.53	295	0.68	0.95	0.19	2.86
	12 and over	4.68	295	1.11	0.90	2.51	6.86
70- to 74-year olds	0–5	84.80	244	2.23	0.97	80.44	89.16
	6–11	9.21	244	1.85	1.00	5.58	12.83
	Under 12	94.01	244	1.52	1.00	91.02	97.00
	12–17	4.05	244	1.19	0.94	1.71	6.39
	18 and over	1.94	244	0.97	1.09	0.04	3.84
	12 and over	5.99	244	1.52	1.00	3.00	8.98
All (60- to 74-year olds)	0–5	80.10	871	1.43	1.05	77.31	82.90
	6–11	11.58	871	1.25	1.16	9.12	14.03
	Under 12	91.68	871	0.98	1.04	89.76	93.59
	12–17	4.41	871	0.72	1.04	2.99	5.83
	18 and over	3.91	871	0.71	1.08	2.52	5.30
	12 and over	8.32	871	0.98	1.04	6.41	10.24

Table continued on page 98.

Table A.1 *(continued)* **Standard errors and 95% confidence intervals for the distribution of CIS-R scores by age (Table 2.1)**

Base	Characteristic	% (p) (adj)	Sample size	True standard error of p	Deft	95% confidence interval LL	UL
All adults							
60- to 64-year olds	0–5	72.62	739	1.78	1.08	69.14	76.11
	6–11	14.28	739	1.38	1.07	11.57	16.99
	Under 12	86.90	739	1.24	1.00	84.47	89.33
	12–17	6.42	739	0.93	1.03	4.60	8.24
	18 and over	6.68	739	0.91	0.99	4.89	8.46
	12 and over	13.10	739	1.24	1.00	10.67	15.53
65- to 69-year olds	0–5	78.01	668	1.60	1.00	74.86	81.15
	6–11	13.29	668	1.34	1.02	10.66	15.91
	Under 12	91.29	668	1.17	1.07	88.99	93.59
	12–17	5.04	668	0.89	1.05	3.30	6.78
	18 and over	3.66	668	0.81	1.11	2.08	5.25
	12 and over	8.71	668	1.17	1.07	6.41	11.01
70- to 74-year olds	0–5	77.97	600	1.62	0.95	74.80	81.13
	6–11	13.60	600	1.43	1.02	10.79	16.41
	Under 12	91.57	600	1.13	0.99	89.36	93.78
	12–17	5.42	600	0.98	1.06	3.51	7.33
	18 and over	3.01	600	0.66	0.95	1.72	4.31
	12 and over	8.43	600	1.13	0.99	6.22	10.64
All (60- to 74-year olds)	0–5	75.93	2007	1.01	1.06	73.96	77.91
	6–11	13.75	2007	0.86	1.11	12.07	15.43
	Under 12	89.68	2007	0.72	1.06	88.28	91.09
	12–17	5.67	2007	0.55	1.06	4.60	6.75
	18 and over	4.65	2007	0.47	0.99	3.73	5.56
	12 and over	10.32	2007	0.72	1.06	8.91	11.72

Table A.2 Standard errors and 95% confidence intervals for the proportion with CIS-R scores of 12 or above by lifetime experience of stressful life events and sex (Table 3.2)

Base	Characteristic	% (p) (adj)	Sample size	True standard error of p	Deft	95% confidence interval LL	UL
Women							
No or 1 event	CIS-R score 12+	4.27	92	1.93	0.91	0.48	8.05
2 events	CIS-R score 12+	9.08	244	1.99	1.08	5.19	12.98
3 events	CIS-R score 12+	11.12	324	1.93	1.10	7.34	14.89
4 events	CIS-R score 12+	12.69	237	2.31	1.07	8.16	17.21
5 events	CIS-R score 12+	13.60	136	3.03	1.03	7.66	19.54
6 and over	CIS-R score 12+	28.77	91	4.54	0.95	19.88	37.66
Men							
No or 1 event	CIS-R score 12+	5.11	54	3.41	1.13	-1.58	11.81
2 events	CIS-R score 12+	2.45	137	1.27	0.96	-0.04	4.94
3 events	CIS-R score 12+	7.91	214	1.93	1.04	4.13	11.69
4 events	CIS-R score 12+	5.77	169	1.99	1.11	1.87	9.68
5 events	CIS-R score 12+	12.33	119	3.20	1.06	6.07	18.59
6 and over	CIS-R score 12+	15.33	165	2.83	1.01	9.77	20.88
All adults							
No or 1 event	CIS-R score 12+	4.62	146	1.80	1.03	1.08	8.15
2 events	CIS-R score 12+	6.43	381	1.32	1.05	3.84	9.01
3 events	CIS-R score 12+	9.74	538	1.35	1.05	7.09	12.38
4 events	CIS-R score 12+	9.46	406	1.62	1.12	6.28	12.64
5 events	CIS-R score 12+	12.91	255	2.22	1.05	8.57	17.26
6 and over	CIS-R score 12+	19.70	256	2.43	0.98	14.93	24.47

Table A.3 Standard errors and 95% confidence intervals for the proportion with ADL difficulties by CIS-R scores at different levels of physical health problems (Table 4.5)

Base	Characteristic	% (p) (adj)	Sample size	True standard error of p	Deft	95% confidence interval LL	UL
No organ systems affected by physical health problems							
CIS-R score <12	No ADL difficulties	85.34	641	1.52	1.09	82.36	88.31
CIS-R score <12	1 ADL difficulties	10.98	641	1.36	1.10	8.31	13.64
CIS-R score <12	2 ADL difficulties	2.44	641	0.66	1.09	1.14	3.73
CIS-R score <12	3 ADL difficulties	0.76	641	0.36	1.04	0.06	1.45
CIS-R score <12	4+ ADL difficulties	0.49	641	0.29	1.04	-0.07	1.06
CIS-R score <12	Any ADL difficulties	14.66	641	1.52	1.09	11.69	17.64
CIS-R score 12+	No ADL difficulties	79.56	26	7.80	0.97	64.28	94.84
CIS-R score 12+	1 ADL difficulties	7.76	26	5.75	1.07	-3.51	19.04
CIS-R score 12+	2 ADL difficulties	5.57	26	4.00	0.87	-2.28	13.41
CIS-R score 12+	3 ADL difficulties	-	26	..	..	..	..
CIS-R score 12+	4+ ADL difficulties	7.11	26	3.88	0.75	-0.49	14.71
CIS-R score 12+	Any ADL difficulties	20.44	26	7.80	0.97	5.16	35.72
1 organ system affected by physical health problems							
CIS-R score <12	No ADL difficulties	64.53	623	1.93	1.01	60.74	68.32
CIS-R score <12	1 ADL difficulties	19.61	623	1.60	1.00	16.48	22.73
CIS-R score <12	2 ADL difficulties	6.30	623	0.90	0.92	4.54	8.06
CIS-R score <12	3 ADL difficulties	4.15	623	0.87	1.09	2.44	5.85
CIS-R score <12	4+ ADL difficulties	5.42	623	1.03	1.14	3.40	7.44
CIS-R score <12	Any ADL difficulties	35.47	623	1.93	1.01	31.68	39.26
CIS-R score 12+	No ADL difficulties	33.81	77	5.60	1.03	22.83	44.79
CIS-R score 12+	1 ADL difficulties	17.99	77	4.70	1.07	8.77	27.21
CIS-R score 12+	2 ADL difficulties	7.38	77	3.25	1.08	1.00	13.75
CIS-R score 12+	3 ADL difficulties	10.89	77	3.63	1.02	3.77	18.01
CIS-R score 12+	4+ ADL difficulties	29.93	77	5.47	1.04	19.20	40.66
CIS-R score 12+	Any ADL difficulties	66.19	77	5.60	1.03	55.21	77.17
2 organ systems affected by physical health problems							
CIS-R score <12	No ADL difficulties	51.52	334	3.06	1.12	45.53	57.51
CIS-R score <12	1 ADL difficulties	25.45	334	2.40	1.00	20.75	30.14
CIS-R score <12	2 ADL difficulties	9.89	334	1.49	0.91	6.96	12.81
CIS-R score <12	3 ADL difficulties	7.60	334	1.44	0.99	4.79	10.42
CIS-R score <12	4+ ADL difficulties	5.54	334	1.20	0.96	3.19	7.89
CIS-R score <12	Any ADL difficulties	48.48	334	3.06	1.12	42.49	54.47
CIS-R score 12+	No ADL difficulties	13.28	61	4.41	1.01	4.63	21.93
CIS-R score 12+	1 ADL difficulties	16.24	61	4.19	0.88	8.03	24.44
CIS-R score 12+	2 ADL difficulties	20.87	61	6.25	1.19	8.62	33.12
CIS-R score 12+	3 ADL difficulties	21.12	61	5.82	1.10	9.71	32.53
CIS-R score 12+	4+ ADL difficulties	28.48	61	6.57	1.13	15.60	41.37
CIS-R score 12+	Any ADL difficulties	86.72	61	4.41	1.01	78.07	95.37

Table continued on page 101.

Table A.3 *(continued)* Standard errors and 95% confidence intervals for the proportion with ADL difficulties by CIS-R scores at different levels of physical health problems (Table 4.5)

Base	Characteristic	% (p) (adj)	Sample size	True standard error of p	Deft	95% confidence interval LL	UL
3+ organ systems affected by physical health problems							
CIS-R score <12	No ADL difficulties	34.90	165	4.31	1.16	26.45	43.34
CIS-R score <12	1 ADL difficulties	20.52	165	3.13	0.99	14.38	26.66
CIS-R score <12	2 ADL difficulties	13.39	165	2.75	1.03	8.00	18.78
CIS-R score <12	3 ADL difficulties	15.60	165	3.41	1.20	8.92	22.28
CIS-R score <12	4+ ADL difficulties	15.59	165	2.95	1.04	9.82	21.37
CIS-R score <12	Any ADL difficulties	65.10	165	4.31	1.16	56.66	73.55
CIS-R score 12+	No ADL difficulties	6.03	50	3.61	1.06	-1.05	13.12
CIS-R score 12+	1 ADL difficulties	21.45	50	6.56	1.12	8.60	34.30
CIS-R score 12+	2 ADL difficulties	5.24	50	2.70	0.85	-0.06	10.54
CIS-R score 12+	3 ADL difficulties	26.46	50	7.30	1.16	12.14	40.77
CIS-R score 12+	4+ ADL difficulties	40.82	50	7.79	1.11	25.55	56.10
CIS-R score 12+	Any ADL difficulties	93.97	50	3.61	1.06	86.88	101.05
All adults							
CIS-R score <12	No ADL difficulties	66.94	1766	1.26	1.12	64.48	69.40
CIS-R score <12	1 ADL difficulties	17.59	1766	1.02	1.13	15.58	19.59
CIS-R score <12	2 ADL difficulties	6.25	1766	0.59	1.02	5.10	7.40
CIS-R score <12	3 ADL difficulties	4.63	1766	0.58	1.15	3.50	5.76
CIS-R score <12	4+ ADL difficulties	4.60	1766	0.52	1.04	3.59	5.62
CIS-R score <12	Any ADL difficulties	33.06	1766	1.26	1.12	30.60	35.52
CIS-R score 12+	No ADL difficulties	26.68	214	3.19	1.05	20.42	32.93
CIS-R score 12+	1 ADL difficulties	17.21	214	2.63	1.02	12.06	22.35
CIS-R score 12+	2 ADL difficulties	10.24	214	2.22	1.07	5.88	14.60
CIS-R score 12+	3 ADL difficulties	16.24	214	2.71	1.07	10.94	21.55
CIS-R score 12+	4+ ADL difficulties	29.63	214	3.42	1.09	22.93	36.33
CIS-R score 12+	Any ADL difficulties	73.32	214	3.19	1.05	67.07	79.58

Table A.4 Standard errors and 95% confidence intervals for the proportion of people with cognitive impairment (TICS-m<21)

by age, marital status and sex (Table 5.1 and 5.2)

Base	Characteristic	% (p) (adj)	Sample size	True standard error of p	Deft	95% confidence interval LL	UL
Women							
60- to 64- year olds	TICS-m score <21	14.46	404	1.70	0.97	11.12	17.80
65- to 69- year olds	TICS-m score <21	22.02	370	2.53	1.17	17.06	26.98
70- to 74- year olds	TICS-m score <21	25.84	355	2.58	1.11	20.78	30.90
Married	TICS-m score <21	19.05	587	1.97	1.21	15.19	22.90
Divorced/separated	TICS-m score <21	18.40	132	4.12	1.22	10.32	26.48
Single	TICS-m score <21	14.22	67	4.45	1.04	5.49	22.95
Widowed	TICS-m score <21	25.79	343	2.26	0.95	21.36	30.21
All women (60-74 years)	TICS-m score <21	19.91	1147	1.42	1.20	17.14	22.68
Men							
60- to 64- year olds	TICS-m score <21	15.86	329	2.11	1.04	11.73	19.99
65- to 69- year olds	TICS-m score <21	16.39	290	2.35	1.08	11.78	21.00
70- to 74- year olds	TICS-m score <21	24.82	241	2.93	1.05	19.07	30.57
Married	TICS-m score <21	18.02	571	1.56	0.97	14.97	21.08
Divorced/separated	TICS-m score <21	15.86	104	3.97	0.10	8.08	23.63
Single	TICS-m score <21	19.52	71	4.57	0.96	10.56	28.48
Widowed	TICS-m score <21	22.96	114	4.20	1.06	14.74	31.18
All men (60-74 years)	TICS-m score <21	17.99	878	1.28	0.98	15.49	20.50
All adults							
60- to 64- year olds	TICS-m score <21	15.15	733	1.34	1.01	12.52	17.78
65- to 69- year olds	TICS-m score <21	19.41	660	1.82	1.18	15.85	22.97
70- to 74- year olds	TICS-m score <21	25.37	596	1.89	1.06	21.66	29.07
Married	TICS-m score <21	18.51	1158	1.25	1.10	16.05	20.97
Divorced/separated	TICS-m score <21	17.25	236	2.93	0.19	11.51	23.00
Single	TICS-m score <21	16.91	138	3.42	1.07	10.20	23.62
Widowed	TICS-m score <21	25.05	457	1.95	0.96	21.22	28.88
All (60- to 74-year olds)	TICS-m score <21	19.18	2007	0.99	1.12	17.25	21.11

B Appendix B The assessment of common mental disorders and cognitive function

B1 Calculation of CIS-R symptom scores and algorithms to produce ICD-10 classification of neurotic disorders

1. Calculation of CIS-R symptom scores

Calculation of symptom score for Somatic Symptoms
Score one for each of:
- Noticed ache or pain/discomfort for four days or more in past seven days.
- Ache or pain/discomfort lasted more than three hours on any day in past week/on that day.
- Ache or pain/discomfort has been very unpleasant in past week.
- Ache or pain/discomfort has bothered you when you were doing something interesting in past week.

Calculation of symptom score for Fatigue
Score one for each of:
- Felt tired/lacking in energy for four days or more in past seven days.
- Felt tired for more than three hours in total on any day in past week.
- Felt so tired/lacking in energy that you've had to push yourself to get things done on at least one occasion during past week.
- Felt tired/lacking in energy when doing things you enjoy or used to enjoy at least once during past week.

Calculation of symptom score for Concentration and forgetfulness
Score one for each of:
- Noticed problems with concentration/memory for four days or more in past week.
- Could not always concentrate on a TV programme, read a newspaper article or talk to someone without mind wandering in past week.
- Problems with concentration actually stopped you from getting on with things you used to do or would like to do.
- Forgot something important in past seven days.

Calculation of symptom score for Sleep Problems
Score one for each of:
- Had problems with sleep for four nights or more out of past seven nights.
- Spent at least $^1/_4$ hour trying to get to sleep on the night with least sleep in the past week.
- Spent three or more hours trying to get to sleep on four nights or more in past week.
- Slept for at least $^1/_4$ hour longer than usual sleeping on the night you slept the longest in the past week.
- Slept for more than three hours longer than usual on four nights or more in past week.

Calculation of symptom score for Irritability
Score one for each of:
- Felt irritable or short tempered/angry on four days or more.
- Felt irritable or short tempered/angry for more than one hour on any day in past week.
- Felt so irritable or short tempered/angry that you wanted to shout at someone in past week (even if you hadn't actually shouted).
- Had arguments, rows or quarrels or lost your temper with someone in past seven days and felt it was unjustified on at least one occasion.

Calculation of symptom score for Worry about Physical Health
Score one for each of:
- Worried about physical health/serious physical illness on four days or more in past seven days.
- Felt that you had been worrying too much, in view of actual health.
- Worrying had been very unpleasant in past week.
- Not able to take mind off health worries at least once by doing something else in past week.

Calculation of symptom score for Depression
Score one for each of:
- Unable to enjoy or take an interest in things as much as usual in past week.
- Felt sad, miserable or depressed/unable to enjoy or take an interest in things on four days or more in past week .

- Felt sad, miserable or depressed/unable to enjoy or take an interest in things for more than three hours in total on any day in past week.
- When sad, miserable or depressed did not become happier when something nice happened, or when in company.

Calculation of symptom score for Depressive ideas

Score one for each of:

- Felt guilty or blamed yourself when things went wrong when it hasn't been your fault at least once in past seven days.
- Felt that you are not as good as other people during past week.
- Felt hopeless, for instance about your future, during past seven days.
- Felt that life isn't worth living in past week.
- Thought of killing yourself in past week.

Calculation of symptom score for Worry

Score one for each of:

- Been worrying about things other than physical health on four or more days out of past seven days.
- Have been worrying too much in view of your circumstances.
- Worrying has been very unpleasant in past week.
- Have worried for more than three hours in total on any of past seven days.

Calculation of symptom score for Anxiety

Score one for each of:

- Felt **generally** anxious/nervous/tense on four or more of past seven days.
- Anxiety/nervousness/tension has been very unpleasant in past week.
- When anxious/nervous/tense, have had one or more of following symptoms:
  – heart racing or pounding
  – hands sweating or shaking
  – feeling dizzy
  – difficulty getting your breath
  – butterflies in your stomach
  – dry mouth
  – nausea or feeling as though you wanted to vomit
- Felt anxious/nervous/tense for more than three hours in total in any one of past seven days.

Calculation of symptom score for Phobias

Score one for each of:

- Felt nervous/anxious about (situation/thing) four or more times in past seven days.
- On occasions when felt anxious/nervous/tense, had one or more of following symptoms:
  – heart racing or pounding
  – hands sweating or shaking
  – feeling dizzy
  – difficulty getting your breath
  – butterflies in your stomach
  – dry mouth
  – nausea or feeling as though you wanted to vomit

- Avoided situation or thing because it would have made you anxious/nervous/tense once or more in past seven days.

Calculation of symptom score for Panic

Score one for each of:

- Anxiety or tension got so bad you got in a panic (eg felt that you might collapse or lose control unless you did something about it) once or more in past week.
- Feelings of panic have been very unpleasant or unbearable in past week.
- This panic/worst of these panics lasted longer than 10 minutes.

Calculation of symptom score for Compulsions

Score one for each of:

- Found yourself doing things over again (that you had already done) on four days or more in last week.
- Have tried to stop repeating behaviour/doing these things over again during past week.
- Repeating behaviour/doing these things over again made you upset or annoyed with yourself in past week.
- Repeated behaviour three or more times during past week.

Calculation of symptom score for Obsessions

Score one for each of:

- Unpleasant thoughts or ideas kept coming into your mind on four days or more in last week.
- Tried to stop thinking any of these thoughts in past week.
- Became upset or annoyed with yourself when you have had these thoughts in past week.
- Longest episode of having such thoughts was $^1/_4$ hour or longer.

2. Algorithms for production of ICD-10 diagnoses of neurosis from the CIS-R ('scores' refer to CIS-R scores)

The neurotic disorders covered in this report were produced from the CIS-R schedule which is described in Chapter 1 and reproduced as part of the questionnaire in the Technical Report of the main survey (*Psychiatric morbidity among adults living in private households, 2000: Technical Report* available on the National Statistics website www.statistics.gov.uk). The production of the six categories of disorder occurred in two stages: first, the informants' responses to the CIS-R were used to produce specific ICD-10 diagnoses of neurosis. This was done by applying the algorithms described below. Second, the range of ICD-10 diagnoses were grouped together to produce categories used in the calculation of prevalence.

F32.00 Mild depressive episode without somatic symptoms

1. Symptom duration ≥ 2 weeks

2. Two or more from:
   - depressed mood
   - loss of interest
   - fatigue

3. Two or three from:
   - reduced concentration
   - reduced self-esteem
   - ideas of guilt
   - pessimism about future
   - suicidal ideas or acts
   - disturbed sleep
   - diminished appetite

4. Social impairment

5. Fewer than four from:
   - lack of normal pleasure /interest
   - loss of normal emotional reactivity
   - a.m. waking ≥ 2 hours early
   - loss of libido
   - diurnal variation in mood
   - diminished appetite
   - loss of ≥ 5% body weight
   - psychomotor agitation
   - psychomotor retardation

F32.01 Mild depressive episode with somatic symptoms

1. Symptom duration ≥ 2 weeks

2. Two or more from:
   - depressed mood
   - loss of interest
   - fatigue

3. Two or three from:
   - reduced concentration
   - reduced self-esteem
   - ideas of guilt
   - pessimism about future
   - suicidal ideas or acts
   - disturbed sleep
   - diminished appetite

4. Social impairment

5. Four or more from:
   - lack of normal pleasure /interest
   - loss of normal emotional reactivity
   - a.m. waking ≥ 2 hours early
   - loss of libido
   - diurnal variation in mood
   - diminished appetite
   - loss of ≥ 5% body weight
   - psychomotor agitation
   - psychomotor retardation

F32.10 Moderate depressive episode without somatic symptoms

1. Symptom duration ≥ 2 weeks

2. Two or more from:
   - depressed mood
   - loss of interest
   - fatigue

3. Four or more from:
   - reduced concentration
   - reduced self-esteem
   - ideas of guilt
   - pessimism about future
   - suicidal ideas or acts
   - disturbed sleep
   - diminished appetite

4. Social impairment

5. Fewer than four from:
   - lack of normal pleasure /interest
   - loss of normal emotional reactivity
   - a.m. waking ≥ 2 hours early
   - loss of libido
   - diurnal variation in mood
   - diminished appetite
   - loss of ≥ 5% body weight
   - psychomotor agitation
   - psychomotor retardation

F32.11 Moderate depressive episode with somatic symptoms

1. Symptom duration ≥ 2 weeks

2. Two or more from:
   - depressed mood
   - loss of interest
   - fatigue

3. Four or more from:
   - reduced concentration
   - reduced self-esteem
   - ideas of guilt
   - pessimism about future
   - suicidal ideas or acts
   - disturbed sleep
   - diminished appetite

4. Social impairment

5. Four or more from:
   - lack of normal pleasure /interest
   - loss of normal emotional reactivity
   - a.m. waking ≥ 2 hours early
   - loss of libido
   - diurnal variation in mood
   - diminished appetite
   - loss of ≥ 5% body weight
   - psychomotor agitation
   - psychomotor retardation

F32.2 Severe depressive episode

1. All three from:
   - depressed mood
   - loss of interest
   - fatigue

2. Four or more from:
   - reduced concentration
   - reduced self-esteem
   - ideas of guilt
   - pessimism about future
   - suicidal ideas or acts
   - disturbed sleep
   - diminished appetite

3. Social impairment

4. Four or more from:
   - lack of normal pleasure /interest
   - loss of normal emotional reactivity
   - a.m. waking ≥ 2 hours early
   - loss of libido
   - diurnal variation in mood
   - diminished appetite
   - loss of ≥ 5% body weight
   - psychomotor agitation
   - psychomotor retardation

F40.00 Agoraphobia without panic disorder

1. Fear of open spaces and related aspects: crowds, distance from home, travelling alone.
2. Social impairment.
3. Avoidant behaviour must be prominent feature.
4. Overall phobia score ≥ 2.
5. No panic attacks.

F40.01 Agoraphobia with panic disorder

1. Fear of open spaces and related aspects: crowds, distance from home, travelling alone.
2. Social impairment.
3. Avoidant behaviour must be prominent feature.
4. Overall phobia score ≥ 2.
5. Panic disorder (overall panic score ≥ 2).

F40.1 Social phobias

1. Fear of scrutiny by other people: eating or speaking in public etc.
2. Social impairment.
3. Avoidant behaviour must be prominent feature.
4. Overall phobia score ≥ 2.

F40.2 Specific (isolated) phobias

1. Fear of specific situations or things, eg animals, insects, heights, blood, flying etc.
2. Social impairment.

3. Avoidant behaviour must be prominent feature.
4. Overall phobia score ≥ 2.

F41.0 Panic disorder

1. Criteria for phobic disorders not met.
2. Recent panic attacks.
3. Anxiety-free between attacks.
4. Overall panic score ≥ 2.

F41.1 Generalised anxiety disorder

1. Duration ≥ 6 months.
2. Free-floating anxiety.
3. Autonomic overactivity.
4. Overall anxiety score ≥ 2.

F41.2 Mixed anxiety and depressive disorder

1. (Sum of scores for each CIS-R section) ≥ 12.
2. Criteria for other categories not met.

F42 Obsessive-compulsive disorder

1. Duration ≥ 2 weeks.
2. At least one act / thought resisted.
3. Social impairment.
4. Overall scores:
   obsession score=4, or
   compulsion score=4, or
   obsession+compulsion scores ≥ 6

3. Grouping neurotic disorders (and depressive episode) into broad categories

Depression

F32.00 and F32.01 were grouped to produce mild depressive episode (i.e. with or without somatic symptoms). F32.10 and F32.11 were similarly grouped to produce Moderate Depressive Episode. Mild depressive episode, moderate depressive episode and Severe depressive episode (F32.2) were then combined to produce the final category of Depression.

Phobias

The ICD-10 phobic diagnoses F40.00, F40.01, F40.1 and F40.2, were combined into one category of phobia.

Categories for analysis

This produced six categories of neurotic for analysis:
- Mixed anxiety and depressive disorder
- Generalised anxiety disorder
- Depressive disorder
- All phobias
- Obsessive compulsive disorder
- Panic disorder

These were further grouped into any anxiety disorders (generalised anxiety disorder, panic, phobia, obsessive compulsive disorder and mixed anxiety and depressive disorder) and any neurotic disorder excluding mixed anxiety and depression.

B2 TICS-m (Telephone Interview for Cognitive Status)

Score '1' for each correct answer and '0' if incorrect

Orientation

1. (i) What day of the week is it?
   (ii) What is today's date? dd/mm/yyyy
   (iii) What season are we in?

2. What is your age?

3. What is your telephone number? (code + number)

Registration/Free Recall

4. I'm going to read you a list of ten words. Please listen carefully and try to remember them. When I am done, tell me as many as you can in any order. Ready?

Now, tell me all the words you can remember

Cabin
Pipe
Elephant
Chest
Silk
Theatre
Watch
Whip
Pillow
Giant

Attention/Calculation

5. Please take away 7 from 100
   Now continue to take 7 away from what you have left over until I ask you to stop.

6. Please count backwards from 20 to 1 — *No mistakes*

Comprehension, Semantic and recent memory

7. What do people usually use to cut paper? — *Scissors*
8. What is the prickly green plant found in the desert? — *Cactus*
9. Who is the reigning monarch now? *Elizabeth/Queen Elizabeth/Queen Elizabeth 2nd*
10. Who is the prime minister now? — *Correct surname*
11. What is the opposite of East? — *West*

Language/Reception

12. Please say this 'Methodist Episcopal' — *Exactly right*

Delayed Recall

13. Please repeat a list of 10 words I read earlier

Cabin
Pipe
Elephant
Chest
Silk
Theatre
Watch
Whip
Pillow
Giant

MAXIMUM of [39]

B3 National Adult Reading Test (NART)

The words included in the test are:

CHORD	HEIR	PLACEBO
ACHE	RADIX	ABSTEMIOUS
DEPOT	ASSIGNATE	DETENTE
AISLE	HIATUS	IDYLL
BOUQUET	SUBTLE	PUERPERAL
PSALM	PROCREATE	AVER
CAPON	GIST	GAUCHE
DENY	GOUGE	TOPIARY
NAUSEA	SUPERFLUOUS	LEVIATHAN
DEBT	SIMILE	BEATIFY
COURTEOUS	BANAL	PRELATE
RAREFY	QUADRUPED	SIDEREAL
EQUIVOCAL	CELLIST	DEMESNE
NAIVE	FAÇADE	SYNCOPE
CATACOMB	ZEALOT	LABILE
GAOLED	DRACHM	CAMPANILE
THYME	AEON	

C

Appendix C Glossary of survey definitions and terms

Educational level

Educational level was based on the highest educational qualification obtained and was initially grouped as indicated below. For analysis purposes these were subsumed under 3 categories: 'A' level or above (categories 1,2 and 3), GCSE level or equivalent (categories 4,5 and 6), and none

1 Degree or higher degree
NVQ Level 5

2 Teaching qualification
HNC/HND
BRC/TEC Higher
BTEC/SCOTVEC Higher
City and Guilds
Full Technological Certificate
Nursing Qualifications (SRN, SCM, RGN, RM, RHV, Midwife)
NVQ Level 4

3 GCE A levels and AS levels
SCE Higher
ONC/OND/BTEC/TEC/BTEC not higher
City and Guilds Advanced/Final Level
GNVQ (Advanced Level)
NVQ Level 3

4 GCE O level passes (Grade A–C if after 1975)
GCSE (Grades A–C)
CSE Grade 1
SCE Ordinary (Bands A–C)
Standard Grade (Level 1–3)
School Certificate or Matric
City and Guilds Craft/Ordinary Level
GNVQ (Intermediate level)
NVQ Level 2

5 CSE Grades 2–5
GCE O level Grades D and E after 1975
GCSE (Grades D,E,F,G)
SCE Ordinary (Bands D and E)
Standard Grade (Level 4,5)
Clerical or Commercial qualifications
Apprenticeships
NVQ Level 1 and GNVQ (Foundation Level)

6 CSE ungraded

7 No formal qualifications

Household

The standard definition used in most surveys carried out by ONS Social Survey Division, and comparable with the 1991 Census definition of a household, was used in this survey. A household is defined as single person or group of people who have the accommodation as their only or main residence and who either share one meal a day or share the living accommodation. (See E McCrossan *A Handbook for interviewers* (1991), HMSO: London.)

Marital Status

Informants were categorised based on their legal marital status.

Neurotic disorders, depression or anxiety disorders

These are characterised by a variety of symptoms such as fatigue and sleep problems, forgetfulness and concentration difficulties, irritability, worry, panic, hopelessness, and obsessions and compulsions, which are present to such a degree that they cause problems with daily activities and distress. The prevalence of neurotic symptoms in the week prior to interview was assessed using the revised version of the Clinical Interview Schedule (CIS-R). A score of 12 or more indicates the presence of significant neurotic symptoms while a score of 18 or more indicates symptoms of a level likely to require treatment.

Social Class

Based on the Registrars general's 1991 *Standard Occupational Classification*, Volume 3 OPCS, HMSO: London social class was ascribed on the basis of the informants own occupation. If the informant was unemployed or economically inactive at the time of interview but had previously worked, social class was based on the most recent previous occupation.

The classification used in the tables are as follows:

Descriptive Definition	*Social Class*
Professional	I
Intermediate occupations	II
Skilled occupations – non-manual	III NM
Skilled occupation – manual	III M
Partly-skilled	IV
Unskilled occupations	V

Armed Forces

Social class was not determined where the subject had never worked, or if the subject was a full-time student or where occupation was inadequately described.

Tenure

Four tenure categories were created:

'Owned outright' means bought without a mortgage or loan or with a mortgage or loan which has been paid off.

'Owned with mortgage' includes co-ownership and shared ownership schemes.

'Rent from LA/HA' means rented from local authorities, New Town corporations or commissions or Scottish Homes, and housing associations which include co-operatives and property owned by charitable trusts.

'Rent from other source' includes rent from organisations (property company, employer or other organisation) and from individuals (relative, friend, employer or other individual).